SAT Subject Test:

MATHEMATICS LEVEL 1

Tenth Edition

RELATED KAPLAN TITLES FOR COLLEGE-BOUND STUDENTS

AP Biology
AP Calculus AB & BC
AP Chemistry
AP English Language & Composition
AP English Literature & Composition
AP Environmental Science
AP European History
AP Human Geography
AP Macroeconomics/Microeconomics
AP Physics B & C
AP Psychology
AP Statistics
AP U.S. Government & Politics
AP U.S. History
AP World History

ACT Strategies, Practice, and Review
ACT Premier
8 Practice Tests for the ACT

SAT Strategies, Practice, and Review
SAT Premier
SAT Total Prep
8 Practice Tests for the SAT
Evidence-Based Reading, Writing, and Essay Workbook for the SAT
Math Workbook for the SAT

SAT Subject Test: Biology E/M
SAT Subject Test: Chemistry
SAT Subject Test: Literature
SAT Subject Test: Mathematics Level 2
SAT Subject Test: Physics
SAT Subject Test: U.S. History

SAT® Subject Test:

MATHEMATICS LEVEL 1

Tenth Edition

KAPLAN) PUBLISHING

New York

SAT® is a trademark registered and/or owned by the College Board, which was not involved in the production of, and does not endorse, this product.

This publication is designed to provide accurate and authoritative information in regard to the subject matter covered. It is sold with the understanding that the publisher is not engaged in rendering legal, accounting, or other professional service. If legal advice or other expert assistance is required, the services of a competent professional should be sought.

© 2017 by Kaplan, Inc.

Published by Kaplan Publishing, a division of Kaplan, Inc.
750 Third Avenue
New York, NY 10017

Printed in the United States of America

10 9 8 7 6 5 4 3 2 1

ISBN 13: 978-1-5062-0922-7

Kaplan Publishing books are available at special quantity discounts to use for sales promotions, employee premiums, or educational purposes. For more information or to purchase books, please call the Simon & Schuster special sales department at 866-506-1949.

Table of Contents

PART THREE: Practice Tests

AVAILABLE ONLINE

FOR ANY TEST CHANGES OR LATE-BREAKING DEVELOPMENTS
kaptest.com/publishing

The material in this book is up-to-date at the time of publication. However, the College Board and Educational Testing Service (ETS) may have instituted changes in the test after this book was published. Be sure to read the materials you receive when you register for the test.

If there are any important late-breaking developments—or any changes or corrections to the Kaplan test preparation materials in this book—we will post that information online at **kaptest.com/publishing**.

For customer service, please contact us at **booksupport@kaplan.com**.

Part One

The Basics

Chapter 1: **Getting Ready for SAT Subject Test: Mathematics**

- Understand the SAT subject tests
- Content of the SAT Subject Test: Mathematics
- Finding your level
- Level of difficulty and scoring

You're serious about going to the college of your choice. You wouldn't have opened this book otherwise. You've made a wise choice, because this book can help you to achieve your goal. It'll show you how to score your best on the SAT Subject Test: Mathematics. But before turning to the math content, let's look at the SAT subject tests generally.

UNDERSTAND THE SAT SUBJECT TESTS

The following background information about the SAT subject test is important to keep in mind as you get ready to prep for the SAT Subject Test: Mathematics.

What Are the SAT Subject Tests?

Known until 1994 as the College Board Achievement Tests and until 2004 as the SAT IIs, the SAT Subject Tests focus on specific disciplines: English, U.S. History, World History, Mathematics, Physics, Chemistry, Biology, and many foreign languages. Each test lasts one hour and consists entirely of multiple-choice questions. On any one test date, you can take one, two, or three subject tests.

How Do the SAT Subject Tests Differ from the SAT?

The SAT is largely a test of verbal and math skills. True, you need to know some vocabulary and some formulas for the SAT, but it's designed to measure how well you read and think rather than how much you remember. The SAT subject tests are very different. They're designed to measure what you know about specific disciplines. Sure, critical reading and thinking skills play a part on these tests, but their main purpose is to determine exactly what you know about math, history, chemistry, and so on.

How Do Colleges Use the SAT Subject Tests?

Many people will tell you that the SAT and SAT subject tests measure only your ability to perform on standardized exams—that they measure neither your reading and thinking skills nor your level of knowledge. Maybe they're right. But these people don't work for colleges. Those schools that require SATs feel that they are an important indicator of your ability to succeed in college. Specifically, they use your scores in one or both of two ways: to help them make admissions and/or placement decisions.

Like the SAT, the SAT subject tests provide schools with a standard measure of academic performance, which they use to compare you to applicants from different high schools and different educational backgrounds. This information helps them to decide if you're ready to handle their curriculum.

SAT subject test scores may also be used to decide what course of study is appropriate for you once you've been admitted. A high score on an SAT Subject Test: Mathematics Level 1 may mean that you'll be exempted from an introductory math course.

> **DUAL ROLE**
>
> Colleges use your SAT subject test scores in both admissions and placement decisions.

Which SAT Subject Tests Should I Take?

The simple answer is: those that you'll do well on. High scores, after all, can only help your chances for admission. Unfortunately, many colleges demand that you take particular tests, usually including one of the Mathematics tests. Some schools will give you some choice in the matter, especially if they want you to take a total of three tests. Before you register to take any tests, therefore, check with the colleges you're interested in to find out exactly which tests they require. Don't rely on high school guidance counselors or admissions handbooks for this information. They might not give you accurate or current information.

> **CALL YOUR COLLEGES**
>
> Many colleges require you to take certain SAT subject tests. Check with all of the schools you're interested in applying to before deciding which tests to take.

When Are the SAT Subject Tests Administered?

Most of the SAT subject tests are administered six times a year: in October, November, December, January, May, and June. A few of the tests are offered less frequently. Due to admissions deadlines, many colleges insist that you take SAT subject tests no later than December or January of your senior year in high school. You may even have to take them sooner if you're interested in applying for "early admission" to a school. Those schools that use scores for placement decisions only may allow you to take SAT subject tests as late as May or June of your senior year. You should check with colleges to find out which test dates are most appropriate for you.

> **COUNT TO THREE**
>
> You can take up to three SAT subject tests in one day.

How Do I Register for SAT Subject Tests?

The College Board administers the SAT subject tests, so you must sign up with them. The easiest way to register is online. Visit the College Board's website at www.collegeboard.org for registration information. If you register online, you immediately get to choose your test date and test center, and you have 24-hour access to print your admission ticket. You'll need access to a credit card to complete online registration.

If you would prefer to register by mail, you must obtain a copy of the *Student Registration Guide for the SAT and SAT Subject Tests*. This publication contains all of the necessary information, including current test dates and fees. It can be obtained at any high school guidance office or directly from the College Board. If you have previously registered for an SAT or an SAT Subject Test, you can reregister by telephone for an additional fee ($15 at the time of this printing). If you choose this option, you should still read the College Board publications carefully before you make any decisions.

How Are the SAT Subject Tests Scored?

The SAT subject tests are scored on a 200–800 scale.

What's a "Good" Score?

That's tricky. The obvious answer is: the score that the colleges of your choice demand. Keep in mind, though, that SAT subject test scores are just one piece of information that colleges will use to evaluate you. The decision to accept or reject you will be based on many criteria, including your high school transcript, your SAT scores, your recommendations, your personal statement, your interview (where applicable), your extracurricular activities, and the like. So failure to achieve the necessary score doesn't automatically mean that your chances of getting in have been damaged. If you really want a numerical benchmark, a score of 600 is considered very solid.

What Should I Bring to the SAT Subject Test?

It's a good idea to get your test materials together the day before the tests. You'll need an admission ticket, a form of identification (check the *Registration Guide* or College Board website to find out what is and what is not permissible), a few sharpened No. 2 pencils, a good eraser, and an approved calculator. Also, make sure that you know how to get to the test center.

CONTENT OF THE SAT SUBJECT TEST: MATHEMATICS

There's a lot of overlap between what's tested on Level 1 and what's tested on Level 2. But there's also a lot that's tested on Level 2 only, and even some math that's tested on Level 1 only.

Level 1 is meant to cover the math you'd get in two years of algebra and one year of geometry. Level 2 is meant to cover that much math plus what you'd get in a year of trigonometry and/or precalculus. There is no calculus on either test.

In order to make room for more questions on more advanced topics, Level 2 has fewer questions on the more basic topics. In fact, it has no plane geometry questions at all. While we've included the official breakdown here, please visit the College Board's website for additional information regarding how the two tests differ in topic area.

CONTACT THE TEST MAKERS

Want to register for the SAT subject tests or get more info? You can get copies of the *Student Registration Guide for the SAT and SAT Subject Tests* from the College Board. If you have a credit card, you can also register for the SAT subject test online. You can register by phone *only* if you have registered for an SAT or SAT subject test in the past.

College Board SAT Program
Domestic: 866-756-7346
International: 212-713-7789
www.collegeboard.org

Approximate Percentage of Content Coverage by Topic

Topic	Approx. % Level 1	Approx. % Level 2
Number and Operations	10–14%	10–14%
Algebra and Functions	38–42%	48–52%
Geometry and Measurement	38–42%	28–32%
Data Analysis, Statistics, and Probability	8–12%	8–12%

CONTENT AT A GLANCE

Level 1 covers two years of algebra and one year of geometry. Level 2 covers two years of algebra, one year of geometry, and one year of trigonometry and/or precalculus. There is no calculus on either test.

FIRM UP THE FOUNDATIONS

Don't review math haphazardly. Start with the fundamentals and work your way up to more advanced and esoteric topics.

Level 2 is weighted toward the more advanced topics, but it still tests your understanding of the basics. For example, Level 2 has *no* plane geometry questions. But to do a lot of the more advanced Level 2 questions—solid geometry, coordinate geometry, trigonometry—you have to know all about plane geometry.

The topics listed in the chart are not equally difficult. However, they do overlap. Think about how you learned these subjects. You didn't start with trigonometry or functions, did you? Of course not. Math is cumulative. Advanced subjects are built upon basic subjects. Firm up the foundations, and work your way up to more advanced topics.

The emphasis in Level 1 is on the foundations, while in Level 2 it's more on the advanced topics. But because the more advanced topics are built upon the basics, it can be said that for Level 2 you need to know everything that's tested on Level 1, plus a lot more.

FINDING YOUR LEVEL

The first thing to do to get ready for SAT Subject Test: Mathematics Level 1 is to be sure you are taking the right test. The information you need to make that decision, besides the differences in content, are level of difficulty, scoring, and reputation.

Level of Difficulty and Scoring

After content, the second and third factors to consider in deciding which test to take are *level of difficulty* and *scoring*. Level 2 questions are considerably more difficult than Level 1 questions. Some Level 2 questions are more difficult because they test more advanced topics. But even the Level 2 questions on basic math are generally more difficult than their counterparts on Level 1. This big difference in level of difficulty, however, is partially offset by differences in the score conversion tables. On Level 1, you would probably need to answer every question correctly to get an 800. On Level 2, however, you can get six or seven questions wrong and still get an 800. On Level 1, you would need a raw score of more than 20 (out of 50) to get a 500, but on Level 2, you can get a 500 with a raw score as low as about 11.

You don't need as many right answers to achieve a particular score on Level 2, so don't assume that you'll get a higher score by taking Level 1. If you've had a year of trigonometry and/or precalculus, you might actually find it easier to reach a particular score goal by taking Level 2.

Reputation

The final factor to consider is reputation. Admissions people know how much more math you have to know to get a good score on Level 2 than on Level 1. Your purpose is to demonstrate how much you've learned in high school. If you've learned enough math to take Level 2, then show it off!

> ### YOU DON'T NEED TO BE PERFECT
>
> On Level 2 you can leave several questions unanswered, or even get them wrong, and still get an 800.

> ### TEST YOUR BEST
>
> If you have the background to take Level 2, don't jump to the conclusion that you'll get a higher score by taking Level 1 instead.

Chapter 2: **SAT Subject Test Mastery**

- Use the structure of the test to your advantage
- Approaching SAT subject test questions
- Work strategically
- Stress management
- The final countdown

Now that you know a little about the SAT subject tests, it's time to let you in on a few basic test-taking skills and strategies that can improve your performance on them. You should practice these skills and strategies as you prepare for these tests.

USE THE STRUCTURE OF THE TEST TO YOUR ADVANTAGE

The SAT subject tests are different from the tests that you're used to taking. On your high school tests, you probably go through the questions in order. You probably spend more time on hard questions than on easy ones, since hard questions are generally worth more points. And you often show your work since your teachers tell you that how you approach questions is as important as getting the right answers.

None of this applies to the SAT subject tests. You can benefit from moving around within the tests, hard questions are worth the same as easy ones, and it doesn't matter how you calculate the answers—only what your answers are.

Take Advantage of the So-Called Guessing Penalty

You might have heard it said that the SAT subject test has a "guessing penalty." That's a misnomer. It's really a *wrong-answer penalty*. If you guess wrong, you get a small penalty. If you guess right, you get full credit.

The fact is, if you can eliminate one or more answer choices as definitely wrong, you'll turn the odds in your favor and actually come out ahead by guessing. The fractional points that you lose are meant to offset the points you might get "accidentally" by guessing the correct answer. With practice, however, you'll see that it's often easy to eliminate *several* answer choices on some of the questions.

The Answer Grid Has No Heart

It sounds simple, but it's extremely important: Don't make mistakes filling out your answer grid. When time is short, it's easy to get confused going back and forth between your test booklet and your grid. If you know the answers but misgrid, you won't get the points. Here's how to avoid mistakes.

Always circle the questions you skip. Put a big circle in your test booklet around any question numbers that you skip. When you go back, these questions will be easy to locate. Also, if you accidentally skip a box on the grid, you'll be able to check your grid against your booklet to see where you went wrong.

Always circle the answers you choose. Circling your answers in the test booklet makes it easier to check your grid against your booklet.

Grid five or more answers at once. Don't transfer your answers to the grid after every question. Transfer them after every five questions. That way, you won't keep breaking your concentration to mark the grid. You'll save time and gain accuracy.

The SAT Subject Tests Are Highly Predictable

Because the format and directions of the SAT subject tests remain unchanged from test to test, you can learn the tests' setups in advance. On test day, the various question types on the tests shouldn't be new to you.

One of the easiest things you can do to help your performance on the SAT subject tests is to understand the directions before taking the test. Since the instructions are always the same, there's no reason to waste a lot of time on test day reading them. Learn them beforehand, as you work through this book and the College Board publications.

Order of Difficulty

SAT Subject Test: Mathematics questions are arranged in order of difficulty. The questions generally get harder as you work through different parts of a test. This pattern can work to your benefit. Try to be aware of where you are in a test. Be careful though. A few hard questions may appear early, a few easy ones late.

When working on more basic problems, you can generally trust your first impulse—the obvious answer is likely to be correct. As you get to the end of a test section, you need to be a bit more suspicious. Now the answers probably won't come as quickly and easily—if they do, look again because the obvious answers may be wrong. Watch out for answers that just "look right." They may be distractors—wrong answer choices deliberately meant to entice you.

Move Around

You're allowed to skip around on the SAT subject tests. High scorers know this fact. They move through the tests efficiently. They don't dwell on any one question, even a hard one, until they've tried every question at least once.

When you run into questions that look tough, circle them in your test booklet and skip them for the time being. Go back and try again after you've answered the easier ones if you've got time. After a second look, troublesome questions can turn out to be remarkably simple.

If you've started to answer a question but get confused, quit and go on to the next question. Persistence might pay off in high school, but it usually hurts your SAT subject test scores. Don't spend so much time answering one hard question that you use up three or four questions' worth of time. That'll cost you points, especially if you don't even get the hard question right.

APPROACHING SAT SUBJECT TEST QUESTIONS

Apart from knowing the setup of the SAT subject tests that you'll be taking, you've got to have a system for attacking the questions. You wouldn't travel around an unfamiliar city without a map, and you shouldn't approach any SAT subject test without a plan. What follows is the best method for approaching the questions systematically.

Think First

Think about the questions before you look at the answers. The test makers love to put distractors among the answer choices. Distractors are answers that look like they're correct, but aren't. If you jump right into the answer choices without thinking first about what you're looking for, you're much more likely to fall for one of these traps.

Be a Good Guesser

You already know that the "guessing penalty" can work in your favor. Don't simply skip questions that you can't answer. Spend some time with them in order to see whether you can eliminate any of the answer choices. If you can, it pays for you to guess.

LEAP AHEAD

You should do the questions in the order that's best for you. Don't pass up the opportunity to score easy points by wasting time on hard questions. Skip hard questions until you've gone through every question once. Come back to them later.

THINK FIRST

Always try to think of the answer to a question before you shop among the answer choices. If you've got some idea of what you're looking for, you'll be less likely to be fooled by "trap" choices.

GUESSING RULE

Don't guess unless you can eliminate at least one answer choice. Don't leave a question blank unless you have absolutely no idea about it.

Pace Yourself

The SAT Subject Tests give you a lot of questions in a short period of time. To get through a test, you can't spend too much time on any single question. Keep moving through the test at a good speed. If you run into a hard question, circle it in your test booklet, skip it, and come back to it later if you have time.

Don't spend the same amount of time on every question. Ideally, you should be able to work through the easier questions at a brisk, steady clip and use a little more time on the harder questions. One caution: Don't rush through basic questions just to save time for the harder ones. The basic questions are points in your pocket. All questions are worth the same number of points. Therefore, don't worry about answering the more difficult questions—work methodically through the easier questions and rack up the points! Remember, you don't earn any extra credit for answering hard questions.

> **SPEED LIMIT**
>
> Work quickly on easier questions to leave more time for harder questions. But don't work so quickly that you lose points by making careless errors. And it's okay to leave some questions blank if you have to—you can still get a high score.

WORK STRATEGICALLY

As you'll see throughout this book, there is often more than one way to solve a problem. Be on the lookout for the quicker route to the answer. Making such choices requires awareness of the options and lots of practice.

- You can set up and solve an equation with or without the aid of a calculator (the subject of chapter 3).
- You can Pick Numbers, for example, when answer choices are algebraic expressions. We'll look at several examples of this approach in the "Mathematics Level 1 Review" chapters.
- You can Backsolve, which is often the quickest route to an answer when the question describes an equation and the answer choices are simple numbers. To Backsolve, generally start with choice (B) or (D). That gives you a 2-in-5 chance that a single calculation will give you the answer (as long as you can tell whether a larger or smaller number is desirable). If you try (B), you'll know the right answer (1) if (B) is correct or (2) if it's too high—making (A) correct. If that calculation isn't enough, only one more is needed. Try the middle number of the remaining three, (D). If it's right, it's your answer. If it's too high, (C) is correct, and if it's too low, (E) is.
- You may be able to "eyeball" for the answer. Look at the figure provided; mark it up if that helps. Often, when no figure is provided, just drawing one will make the answer apparent.

Locate quick points. Some questions can be done more quickly than others because they require less work or because choices can be eliminated more easily. If you start to run out of time, look for these quicker questions.

Set a target score. Naturally, you want the best score you can earn to maximize your college options. But choose a realistic target score that is above the average for the school or schools you want to attend.

By keeping an eye on which questions you are sure of and which you guessed on, you can monitor your progress toward this goal. Of course, you shouldn't stop practicing (or taking the test) when you reach your target score—but you can be more relaxed and confident.

When you take an SAT subject test, you have one clear objective in mind: to score as many points as you can. It's that simple. The rest of this book will show you how to do that on the SAT Subject Test: Mathematics Level 1.

STRESS MANAGEMENT

The countdown has begun. Your date with THE TEST is looming on the horizon. Anxiety is on the rise. The butterflies in your stomach have gone ballistic. Perhaps you feel as if the last thing you ate has turned into a lead ball. Your thinking is getting cloudy. Maybe you think you won't be ready. Maybe you already know your stuff, but you're going into panic mode anyway. Worst of all, you're not sure of what to do about it.

Don't freak! It is possible to tame that anxiety and stress—before and during the test. We'll show you how. You won't believe how quickly and easily you can deal with that killer anxiety.

Identify the Sources of Stress

In the space provided, jot down anything you identify as a source of your test-related stress. The idea is to pin down that free-floating anxiety so that you can take control of it. Here are some common examples to get you started:

- I always freeze up on tests.
- I'm nervous about trig (or functions, or geometry, etc.).
- I need a good/great score to go to Acme College.
- My older brother/sister/best friend/girl- or boyfriend did really well. I must at least match their scores.

- My parents, who are paying for school, will be really disappointed if I don't test well.
- I'm afraid of losing my focus and concentration.
- I'm afraid I'm not spending enough time preparing.
- I study like crazy, but nothing seems to stick in my mind.
- I always run out of time and get panicky.
- I feel as though thinking is becoming like wading through thick mud.

Sources of Stress

_____   _____

_____   _____

_____   _____

_____   _____

Take a few minutes to think about the things you've just written down. Then rewrite them in some sort of order. List the statements you most associate with your stress and anxiety first and put the least disturbing items last. Chances are, the top of the list is a fairly accurate description of exactly how you react to test anxiety, both physically and mentally. The later items usually describe your fears (disappointing Mom and Dad, looking bad, etc.). As you write the list, you're forming a hierarchy of items so you can deal first with the anxiety provokers that bug you most. Very often, taking care of the major items from the top of the list goes a long way toward relieving overall testing anxiety. You probably won't have to bother with the stuff you placed last.

> **THINK GOOD THOUGHTS**
>
> Create a set of positive but brief affirmations and mentally repeat them to yourself just before you fall asleep at night. (That's when your mind is very open to suggestion.) You'll find yourself feeling a lot more positive in the morning.
>
> Periodically repeating your affirmations during the day makes them more effective.

Make the Most of Your Prep Time

Lack of control is one of the prime causes of stress. A ton of research shows that if you don't have a sense of control over what's happening in your life, you can easily end up feeling helpless and hopeless. So just having concrete things to do and to think about—taking control—will help reduce your stress.

Strengths and Weaknesses

Take one minute to list the areas of the test that you are good at. They can be general ("algebra") or specific ("quadratic equations"). Put down as many as you can think of, and if possible, time yourself. Write for the entire time; don't stop writing until you've reached the one-minute stopping point.

Strong Test Subjects

_____   _____

_____   _____

_____   _____

_____   _____

Next, take one minute to list areas of the test you're not so good at, just plain bad at, have failed at, or keep failing at. Again, keep it to one minute and continue writing until you reach the cutoff. Don't be afraid to identify and write down your weak spots! In all probability, as you do both lists, you'll find you are strong in some areas and not so strong in others. Taking stock of your assets and liabilities lets you know the areas you don't have to worry about and the ones that will demand extra attention and effort.

Weak Test Subjects

_____   _____

_____   _____

_____   _____

_____   _____

Facing your weak spots gives you some distinct advantages. It helps a lot to find out where you need to spend extra effort. Increased exposure to tough material makes it more familiar and less intimidating. (After all, we mostly fear what we don't know and are probably afraid to face.) You'll feel better about yourself because you're dealing directly with areas of the test that bring on your anxiety. You can't help feeling more confident when you know you're actively strengthening your chances of earning a higher overall test score.

Now, go back to the "good" list and expand it for two minutes. Take the general items on that first list and make them more specific; take the specific items and expand them into more general conclusions. Naturally, if anything new comes to mind, jot it down. Focus all of your attention and effort on your strengths. Don't underestimate yourself or your abilities. Give yourself full credit. At the same time, don't list strengths you don't really have; you'll only be fooling yourself.

Expanding from general to specific might go as follows. If you listed "algebra" as a broad topic you feel strong in, you would then narrow your focus to include areas of this subject about which you are particularly knowledgeable. Your areas of strength might include multiplying polynomials, working with exponents, factoring, solving simultaneous equations, etc.

VERY SUPERSTITIOUS

Stress expert Stephen Sideroff, PhD, tells of a client who always stressed out before, during, and even after taking tests. Yet she always got outstanding scores. It became obvious that she was thinking superstitiously—subconsciously believing that the great scores were a result of her worrying. She didn't trust herself and believed that if she didn't worry, she wouldn't study hard enough. Sideroff convinced her to take a risk and work on relaxing before her next test. She did, and her test results were still as good as ever—which broke her cycle of superstitious thinking.

GET IT TOGETHER

Don't work in a messy or cramped area. Before you sit down to study, clear yourself a nice, open space. And make sure you have books, paper, pencils—whatever tools you will need—within easy reach before you sit down to study.

Whatever you know comfortably goes on your "good" list. Okay. You've got the picture. Now, get ready, check your starting time, and start writing down items on your expanded "good" list.

Strong Test Subjects: An Expanded List

_____   _____

_____   _____

_____   _____

_____   _____

After you've stopped, check your time. Did you find yourself going beyond the two minutes allotted? Did you write down more things than you thought you knew? Is it possible you know more than you've given yourself credit for? Could that mean you've found a number of areas in which you feel strong?

You just took an active step toward helping yourself. Notice any increased feelings of confidence? Enjoy them.

Here's another way to think about your writing exercise. Every area of strength and confidence you can identify is much like having a reserve of solid gold at Fort Knox. You'll be able to draw on your reserves as you need them. You can use your reserves to solve difficult questions, maintain confidence, and keep test stress and anxiety at a distance. The encouraging thing is that every time you recognize another area of strength, succeed at coming up with a solution, or get a good score on a test, you increase your reserves. And there is absolutely no limit to how much self-confidence you can have or how good you can feel about yourself.

Imagine Yourself Succeeding

This next little group of exercises is both physical and mental. It's a natural follow-up to what you've just accomplished with your lists.

First, get yourself into a comfortable sitting position in a quiet setting. Wear loose clothes. If you wear glasses, take them off. Then, close your eyes and breathe in a deep, satisfying breath of air. Really fill your lungs until your rib cage is fully expanded and you can't take in any more. Then, exhale the air completely. Imagine you're blowing out a candle with your last little puff of air. Do this two or three more times, filling your lungs to their maximum and emptying them totally. Keep your eyes closed, comfortably but not tightly. Let your body sink deeper into the chair as you become even more comfortable.

With your eyes shut, you can notice something very interesting. You're no longer dealing with the worrisome stuff going on in the world outside of you. Now you can concentrate on what happens *inside* you. The more you recognize your own physical reactions to stress and anxiety, the more you can do about them. You might not realize it, but you've begun to regain a sense of being in control.

Let images begin to form on the "viewing screens" on the back of your eyelids. You're experiencing visualizations from the place in your mind that makes pictures. Allow the images to come easily and naturally; don't force them. Imagine yourself in a relaxing situation. It might be in a special place you've visited before or one you've read about. It can be a fictional location that you create in your imagination, but a real-life memory of a place or situation you know is usually better. Make it as detailed as possible, and notice as much as you can.

Stay focused on the images as you sink farther back into your chair. Breathe easily and naturally. You might have the sensations of any stress or tension draining from your muscles and flowing downward, out your feet and away from you.

Take a moment to check how you're feeling. Notice how comfortable you've become. Imagine how much easier it would be if you could take the test feeling this relaxed and in this state of ease. You've coupled the images of your special place with sensations of comfort and relaxation. You've also found a way to become relaxed simply by visualizing your own safe, special place.

Now, close your eyes and start remembering a real-life situation in which you did well on a test. If you can't come up with one, remember a situation in which you did something (academic or otherwise) that you were really proud of—a genuine accomplishment. Make the memory as detailed as possible. Think about the sights, the sounds, the smells, even the tastes associated with this remembered experience. Remember how confident you felt as you accomplished your goal. Now start thinking about the upcoming test. Keep your thoughts and feelings in line with that successful experience. Don't make comparisons between them. Just imagine taking the upcoming test with the same feelings of confidence and relaxed control.

This exercise is a great way to bring the test down to earth. You should practice this exercise often, especially when the prospect of taking the exam starts to bum you out. The more you practice it, the more effective the exercise will be for you.

OCEAN DUMPING

Visualize a beautiful beach, with white sand, blue skies, sparkling water, a warm sun, and seagulls. See yourself walking on the beach, carrying a small plastic pail. Stop at a good spot and put your worries and whatever may be bugging you into the pail. Drop it at the water's edge and watch it drift out to sea. When the pail is out of sight, walk on.

COUNSELING

Don't forget that your school probably has counseling available. If you can't conquer test stress on your own, make an appointment at the counseling center. That's what counselors are there for.

Control Physical Stress

Exercise Your Frustrations Away

Whether it is jogging, walking, biking, mild aerobics, pushups, or a pickup basketball game, physical exercise is a very effective way to stimulate both your mind and body and to improve your ability to think and concentrate. A surprising number of students get out of the habit of regular exercise, ironically because they're spending so much time prepping for exams. Also, sedentary people—this is a medical fact—get less oxygen to the blood and hence to the head than active people. You can live fine with a little less oxygen; you just can't think as well.

Any big test is a bit like a race. Thinking clearly at the end is just as important as having a quick mind early on. If you can't sustain your energy level in the last sections of the exam, there's too good a chance you could blow it. You need a fit body that can weather the demands any big exam puts on you. Along with a good diet and adequate sleep, exercise is an important part of keeping yourself in fighting shape and thinking clearly for the long haul.

There's another thing that happens when students don't make exercise an integral part of their test preparation. Like any organism in nature, you operate best if all your "energy systems" are in balance. Studying uses a lot of energy, but it's all mental. When you take a study break, do something active instead of raiding the fridge or vegging out in front of the TV. Take a 5- to 10-minute activity break for every 50 or 60 minutes that you study. The physical exertion gets your body into the act, which helps to keep your mind and body in sync. Then, when you finish studying for the night and hit the sack, you won't lie there, tense and unable to sleep because your head is overtired and your body wants to pump iron or run a marathon.

One warning about exercise, however: It's not a good idea to exercise vigorously right before you go to bed. This could easily cause sleep-onset problems. For the same reason, it's also not a good idea to study right up to bedtime. Make time for a "buffer period" before you go to bed: For 30 to 60 minutes, just take a hot shower, meditate, or simply veg out.

TAKE A HIKE, PAL

When you're in the middle of studying and hit a wall, take a short, brisk walk. Breathe deeply and swing your arms as you walk. Clear your mind. (And don't forget to look for flowers that grow in the cracks of the sidewalk.)

PLAY THE MUSIC

If you want to play music, keep it low and in the background. Music with a regular, mathematical rhythm—reggae, for example—aids the learning process. A recording of ocean waves is also soothing.

Take A Deep Breath

Conscious attention to breathing is an excellent way of managing test stress (or any stress, for that matter). The majority of people who get into trouble during tests take shallow breaths. They breathe using only their upper chests and shoulder muscles and may even hold their breath for long periods of time. Conversely, the test taker who by accident or design keeps breathing normally and rhythmically is likely to be more relaxed and in better control during the entire test experience.

So, now is the time to get into the habit of relaxed breathing. Do the next exercise to learn to breathe in a natural, easy rhythm. By the way, this is another technique you can use during the test to collect your thoughts and ward off excess stress. The entire exercise should take no more than three to five minutes.

With your eyes still closed, breathe in slowly and *deeply* through your nose. Hold the breath for a bit, and then release it through your mouth. The key is to breathe slowly and deeply by using your diaphragm (the big band of muscle that spans your body just above your waist) to draw air in and out naturally and effortlessly. Breathing with your diaphragm encourages relaxation and helps minimize tension. Try it and notice how relaxed and comfortable you feel.

THE FINAL COUNTDOWN

Quick Tips for the Days Just Before the Exam

- The best test takers do less and less as the test approaches. Taper off your study schedule and take it easy on yourself. You want to be relaxed and ready on the day of the test. Give yourself time off, especially the evening before the exam. By then, if you've studied well, everything you need to know is firmly stored in your memory banks.

- Positive self-talk can be extremely liberating and invigorating, especially as the test looms closer. Tell yourself things such as "I choose to take this test," rather than "I have to"; "I will do well," rather than "I hope things go well"; "I can," rather than "I cannot." Be aware of negative, self-defeating thoughts and images and immediately counter any you become aware of. Replace them with affirming statements that encourage your self-esteem and confidence. Create and practice visualizations that build on your positive statements.

- Get your act together sooner rather than later. Have everything (including choice of clothing) laid out days in advance. Most important, know where the test will be held and the easiest, quickest way to get there. You will gain great peace of mind if you know that all the little details—gas in the car, directions, etc.—are firmly in your control before the day of the test.

CYBERSTRESS

If you spend a lot of time in cyberspace anyway, do a search for the phrase *stress management*. There's a ton of stress advice on the web, including material specifically for students.

NUTRITION AND STRESS: THE DOS AND DON'TS

Do eat:

- Fruits and vegetables (raw is best, or just lightly steamed or nuked)
- Low-fat protein such as fish, skinless poultry, beans, and legumes (like lentils)
- Whole grains such as brown rice, whole-wheat bread, and pastas (no bleached flour)

Don't eat:

- Refined sugar; sweet, high-fat snacks (Simple carbohydrates like sugar make stress worse, and fatty foods lower your immunity.)
- Salty foods (They can deplete potassium, which you need for nerve functions.)

THE RELAXATION PARADOX

Forcing relaxation is like asking yourself to flap your arms and fly. You can't do it, and every push and prod only gets you more frustrated. Relaxation is something you don't work at. You simply let it happen. Think about it. When was the last time you tried to force yourself to go to sleep and it worked?

DRESS FOR SUCCESS

On the day of the test, wear loose layers. That way, you'll be prepared no matter what the temperature of the room is. (An uncomfortable temperature will just distract you from the job at hand.)

And, if you have an item of clothing that you tend to feel "lucky" or confident in—a shirt, a pair of jeans, whatever—wear it. A little totem couldn't hurt.

- Experience the test site a few days in advance. This is very helpful if you are especially anxious. If at all possible, find out what room your part of the alphabet is assigned to and try to sit there (by yourself) for a while. Better yet, bring some practice material and do at least a section or two, if not an entire practice test, in that room. In this situation, familiarity doesn't breed contempt; it generates comfort and confidence.

- Forego any practice on the day before the test. It's in your best interest to marshal your physical and psychological resources for 24 hours or so. Even racehorses are kept in the paddock and treated like royalty the day before a race. Keep the upcoming test out of your consciousness; go to a movie, take a pleasant hike, or just relax. Don't eat junk food or tons of sugar. And—of course—get plenty of rest the night before. Just don't go to bed too early. It's hard to fall asleep earlier than you're used to, and you don't want to lie there thinking about the test.

Handling Stress During the Test

The biggest stress monster will be the test itself. Fear not; there are methods of quelling your stress during the test.

- Keep moving forward instead of getting bogged down in a difficult question. You don't have to get everything right to achieve a fine score. The best test takers skip difficult material temporarily in search of the easier stuff. They mark the ones that require extra time and thought. This strategy buys time and builds confidence so you can handle the tough stuff later.

- Don't be thrown if other test takers seem to be working more furiously than you are. Continue to spend your time patiently thinking through your answers; doing so leads to better results. Don't mistake the other people's sheer activity as signs of progress and higher scores.

- *Keep breathing!* Weak test takers tend to forget to breathe properly as the test proceeds. They start holding their breath without realizing it, or they breathe erratically or arrhythmically. Improper breathing interferes with clear thinking.

- Some quick isometrics during the test—especially if concentration is wandering or energy is waning—can help. Try this: Put your palms together and press intensely for a few seconds. Concentrate on the tension you feel through your palms, wrists, forearms, and up into your biceps and shoulders. Then, quickly release the pressure. Feel the difference as you let go. Focus on the warm relaxation that floods through the muscles. Now you're ready to return to the task.

- Here's another isometric that will relieve tension in both your neck and eye muscles. Slowly rotate your head from side to side, turning your head and eyes to look as far back over each shoulder as you can. Feel the muscles stretch on one side of your neck as they contract on the other. Repeat five times in each direction.

With what you've just learned here, you're armed and ready to do battle with the test. This book and your studies will give you the information you'll need to answer the questions. It's all firmly planted in your mind. You also know how to deal with any excess tension that might come along when you're studying for and taking the exam. You've experienced everything you need to tame your test anxiety and stress. You're going to get a great score.

PACK YOUR BAG

Gather your test materials the day before the test. You'll need these:

- Admission ticket
- Proper form of ID
- Some sharpened No. 2 pencils
- Good eraser
- An approved graphing or scientific calculator
- Spare calculator batteries

Chapter 3: **The Calculator**

- Have the right calculator
- Use your calculator strategically

In this chapter we'll take an important look at the calculator. We'll discuss what kind to use and how and when to use it.

HAVE THE RIGHT CALCULATOR

First of all, get a graphing or scientific calculator. SAT Subject Test: Mathematics is not like the SAT, for which a calculator is permitted but not really necessary. On this test, a calculator is essential.

Know what kind of calculator you should use. According to the College Board, the majority of students bring a graphing calculator to the SAT Mathematics tests; therefore, the tests are developed with the expectation that most students are using graphing calculators.

Be sure your calculator performs the following functions:

- Sine, cosine, and tangent
- Arcsine, arccosine, and arctangent
- Squares, cubes, and other powers
- Square roots and cube roots
- Base-10 logarithms

No laptops allowed. Excluded are laptops or other computers, tablets, cell phones, or smartphones; machines that print, make noise, or need to be plugged in; or calculators with a typewriter keypad or an angled readout screen. Check the College Board website for further specifics.

Advantages of a Graphing or Scientific Calculator

In addition to that list of necessary functions, there are other advantages to using a scientific or graphing calculator.

Scientific calculators with a two-line display let users see what they are typing before performing the operation, so they can catch input errors. Most scientific calculators will also have parentheses keys; these calculators will handle PEMDAS—order of operations.

Most scientific calculators offer statistical operations: calculating permutations ($_nP_r$), combinations ($_nC_r$), and factorials (!). Newer calculators also have fraction and mixed number capabilities. Many have two very useful keys—the ANS key and the ENTRY key. The ANS key is a substitute for typing in the last calculated answer—saving time and avoiding errors. The ENTRY key will re-enter the last string of calculator keystrokes. This is handy for repeated calculations or to correct an incorrect keystroke sequence.

In addition to all those features, graphing calculators have graphing features, including the ability to

- display graphs;
- find the intersection of simultaneous functions;
- find the roots of functions;
- display the associated table for a graphed function;
- find the equation for a table of values (regressions); and
- determine maximums and minimums of functions.

Here are a few cautions when using calculators:

- Keep in mind that the graphing calculator graphs functions only. It is imperative that you understand this if you try to graph a circle or an ellipse.
- Don't round any values until the final step in a problem. Rounding too soon could result in an incorrect solution.
- Always double-check the keystrokes that you enter. It is easy to make a mistake. The home screen of a graphing calculator will allow you to see what was input even after the function is executed.

Get to know your calculator. Getting the right calculator is only the first step. You must then get used to working with it. Going into the test with an unfamiliar calculator is not much better than going in with no calculator at all.

Practice using your calculator on testlike questions. It's not enough to know how to use your calculator. You need to know how to use it effectively on SAT Subject Test: Mathematics questions. Whenever you work with this book, whenever you take a practice test, practice with the very calculator you will use on test day. With some

DON'T FORGET PEMDAS

Parentheses
Exponents
Multiplication
Division
Addition
Subtraction

experience you will learn when to use and when *not* to use, how to use and how *not* to use your calculator on test questions.

Make sure your calculator is in good working order. You'll feel a lot more confident if you put new batteries in your calculator the night before the test and then check it out to see that it's working properly. You should also take spare batteries with you to the test.

USE YOUR CALCULATOR STRATEGICALLY

Here are some tips for using your calculator strategically during the test.

Don't Use Your Calculator Too Often

You will not need your calculator for every question. Top scorers use their calculators for 40–50 percent of the questions on Level 1 and for 55–65 percent of the questions on Level 2. This is still a math test, not a calculator test. Success depends more on your problem-solving skills than on your calculator skills. The calculator is just one tool. Use it wisely and sparingly.

Here's an example in which the calculator is no help:

1. If $x \neq \pm 1$, then $\dfrac{x+1}{x-1} - \dfrac{x-1}{x+1} =$

(A) $\dfrac{2x}{x-1}$ (B) $\dfrac{2x}{x^2+1}$ (C) $\dfrac{2x}{x^2-1}$ (D) $\dfrac{4x}{x^2+1}$ (E) $\dfrac{4x}{x^2-1}$

To answer this question, you need to be adept at algebraic manipulation. Your calculator won't help here. To subtract fractions, even algebraic fractions like these, you need a common denominator, which in this case is the product of the denominators $x - 1$ and $x + 1$:

$$\frac{x+1}{x-1} - \frac{x-1}{x+1} = \frac{(x+1)(x+1)}{(x-1)(x+1)} - \frac{(x-1)(x-1)}{(x-1)(x+1)}$$

$$= \frac{x^2+2x+1}{x^2-1} - \frac{x^2-2x+1}{x^2-1}$$

$$= \frac{x^2+2x+1-x^2+2x-1}{x^2-1}$$

$$= \frac{4x}{x^2-1}$$

The answer is (E).

Here's a question for which you have a choice. You can answer it just as quickly and easily with or without a calculator.

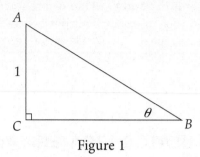

Figure 1

2. In Figure 1, if sin θ = 0.5, what is the length of $\overline{BC}$?

    (A)   1.25
    (B)   1.41
    (C)   1.50
    (D)   1.73
    (E)   2.00

You could use your calculator to find out what angle has a sine of 0.5, but you might just know that it's 30°. This is a 30-60-90 triangle, and $\overline{BC}$ is the longer leg, which is equal to the length of the shorter leg times $\sqrt{3}$. Since AC = 1, then $BC = \sqrt{3}$. You could use your calculator to find the square root of 3, but you might know that it's about 1.73. The answer is (D).

Don't Use Your Calculator Too Early

When you do use your calculator, it will usually be at a later stage in solving a problem. Never start punching calculator keys before you've given the problem a little thought. Know what you're doing and where you're going.

Here's a question that really requires a calculator, but not until a late problem-solving stage:

> **DON'T GET PUNCHY**
>
> Don't start punching calculator keys until you've given the problem some thought.

3. If 0° < x < 90°, and $\tan^2 x - \tan x = 6$, which of the following could be the value of x in degrees (rounded to the nearest degree)?

    (A) 63   (B) 67   (C) 72   (D) 77   (E) 81

What you have here is essentially a quadratic equation in which the unknown is tan x. For simplicity's sake, substitute y for tan x:

$$\tan^2 x - \tan x = 6$$
$$y^2 - y = 6$$

Now solve for y :

$$y^2 - y = 6$$
$$y^2 - y - 6 = 0$$
$$(y-3)(y+2) = 0$$
$$y = 3 \text{ or } -2$$

Now you know that tan x = 3 or –2. Since x is a positive acute angle, the tangent is positive, and tan x = 3. Now's the time to use your calculator. You want to know what acute angle has a tangent of 3. In other words, what you're looking for is the arctangent of 3. Nobody expects you to know the value of arctan(3) off the top of your head. Use your calculator:

$$\tan x = 3$$
$$x = \arctan(3) \approx 71.57$$

The answer is (C).

In the next example, a calculator is not required, but knowledge of certain operations could save you time.

4. Given the table below, which function represents the values in the table?

x	y
1	−1
3	5
5	11
7	17
9	23

   (A) $y = 2x - 3$ (B) $x = 3y - 4$ (C) $x = 2y - 3$ (D) $y = 3x - 4$ (E) $y = x - 2$

One way to solve this problem is to substitute the values of x from the table and see which function results in the corresponding y-values. The correct answer, (D), will give an outcome of each y-value when the x-value to its left is plugged in for x. For example, $3(1) - 4 = 3 - 4 = -1$, so the first pair satisfies the equation. Since $3(3) - 4 = 9 - 4 = 5$, the second pair also satisfies the equation, and so on.

An alternative method would be to input the *x*- and *y*-values into two separate lists on a graphing calculator and perform a linear regression on the data. This approach may prove to be much less time consuming than trial and error; however, you need to be both familiar with and comfortable with the method for finding a linear regression on your particular calculator.

Here is a question in which the calculator is essential.

5. Solve for *x*:
    $7^x = 18.52$

    (A) 0.18   (B) 1.5   (C) 2.65   (D) 21.9   (E) 129.64

$$\log 7^x = \log 18.52$$
$$x (\log 7) = \log 18.52$$
$$x = \frac{\log 18.52}{\log 7}$$
$$x \approx 1.5, \text{ or choice (B)}$$

It is best to use the calculator on the last step and enter both log functions in the same set of keystrokes. If you calculate the log of 18.52 alone, you will get 1.267640982. You could make careless errors and waste precious time as you try to retype this number into the calculator.

Another approach to this problem would be to try each answer choice, using the power key on your keypad.

In the following problem, a graphing calculator would give you a decided advantage.

6. What is a possible solution to the system?

$$y = x^2 + 3x + 2$$
$$x = \frac{1}{2} y - 2$$

    (A) (–1, 2)   (B) (0, 4)   (C) (0, 2)   (D) (1, 6)   (E) (2, 12)

This type of problem can be solved by the substitution method for simultaneous equations. An efficient alternative method is first to solve each equation for *y*, then graph the two functions on a graphing calculator. You can then use the Intersect function to solve the system, arriving at the points (1,6) and (–2,0). Choice (D) names one of these.

To solve the following volume problem, you can use the cube root, or $\sqrt[x]{x}$, key on the calculator. If your calculator does not have these keys, remember that to find a cube root you can use the power key and enter the exponent of $\frac{1}{3}$.

7. What is the edge length of a cube whose volume is 4,096 cm³?

    (A) 16 cm   (B) 64 cm   (C) 455.11 cm   (D) 1,365.33 cm   (E) none of these

$$V = s^3$$
$$4{,}096 = s^3$$
$$\sqrt[3]{4{,}096} = s$$
$$16 \text{ cm} = s, \text{ choice (A)}$$

In any of the given examples, keep in mind that the calculator is a tool for calculating. Your brain is the most important tool you will take into the test. Pay attention as you read further in the lessons to see how and when the calculator should be used. Keep your trusty calculator at your side, but remember to practice restraint.

Part Two

Mathematics Level 1 Review

Chapter 4: **Algebra**

- Expressions and exponents
- Operations with polynomials and factoring
- The Golden Rule of equations
- Quadratic equations

Math 1 includes a lot of algebra. According to the test makers' official breakdown, more than 25 percent of Math 1 questions are algebra questions. But that's counting only the ones that are explicitly and primarily algebra questions. In fact, almost all questions involve algebra. Most functions questions and coordinate geometry questions are algebraic. Most plane geometry and solid geometry questions involve algebraic formulas, and many use algebraic expressions for lengths and angle measures. Most word problems are best solved algebraically. *Algebra is fundamental to Math 1. You can't get a good score without it.*

> **ALGEBRA FACTS AND FORMULAS IN THIS CHAPTER**
>
> - The Five Rules of Exponents
> - Combining Like Terms
> - Multiplying Monomials
> - Multiplying Binomials—FOIL
> - Classic Factorables
> - Quadratic Formula
> - Inequalities and Absolute Value

HOW TO USE THIS CHAPTER

Maybe you know all the algebra you need. Find out by taking the Algebra Diagnostic Test. The ten Diagnostic questions are typical of those on the Mathematics subject tests. Check your answers using the answer key following the test. No matter how you score, don't worry! The answer key also shows where to find a detailed explanation for each question. The "Find Your Study Plan" section that follows the test will suggest next steps based on your performance on the Diagnostic.

Find Your Level

How you use this chapter really depends on how much time you have to prep. Find the level and pace that works best for you.

Standard Plan. No matter how well you do on the Algebra Diagnostic Test, go on and read everything in this chapter and do all the practice problems.

> *Shortcut: If you can answer at least seven of the ten questions on the Algebra Diagnostic Test correctly, then you already know the material in this chapter well enough to move on.*

Panic Plan. The material in this chapter is vital. If you can't get most of the questions on the Algebra Diagnostic Test right, you should study this chapter—even if it's just two days before test day.

ALGEBRA DIAGNOSTIC TEST

10 Questions (12 Minutes)

Directions: Solve the following problems. Fill in the oval corresponding to the best answer choice in the grid to the right of each question. (Answers are on page 38.)

1. If $x = y^2$ and $y = \dfrac{5}{k}$, what is the value of x when $k = \dfrac{1}{2}$?

   (A)   2.50
   (B)   3.16
   (C)   6.25
   (D)   20.00
   (E)   100.00   Ⓐ Ⓑ Ⓒ Ⓓ Ⓔ

2. For all $xyz \neq 0$, $\dfrac{6x^2 y^{12} z^6}{(2x^2 yz)^3} =$

   (A) $\dfrac{y^4 z^2}{x^3}$

   (B) $\dfrac{y^9 z^4}{x^4}$

   (C) $\dfrac{y^9 z^3}{2x^3}$

   (D) $\dfrac{3y^4 z^2}{4x^3}$

   (E) $\dfrac{3y^9 z^3}{4x^4}$ Ⓐ Ⓑ Ⓒ Ⓓ Ⓔ

3. When $2x^3 + 3x^2 - 4x + k$ is divided by $x + 2$, the remainder is 3. What is the value of k ?

   (A) −1   (B) 1   (C) 2   (D) 3   (E) 5

                 Ⓐ Ⓑ Ⓒ Ⓓ Ⓔ

KAPLAN)

4. For all $x \neq \pm 3$, $\dfrac{3x^2 - 11x + 6}{9 - x^2} =$

    (A) $\dfrac{2 - 3x}{x - 3}$

    (B) $\dfrac{2 - 3x}{x + 3}$

    (C) $\dfrac{2x - 3}{x - 3}$

    (D) $\dfrac{3x - 2}{x + 3}$

    (E) $\dfrac{3x - 2}{x - 3}$ Ⓐ Ⓑ Ⓒ Ⓓ Ⓔ

5. If $\sqrt[3]{8x + 6} = -3$, what is the value of x ?

    (A) -4.125

    (B) -2.625

    (C) -1.875

    (D) -1.125

    (E) 2.625 Ⓐ Ⓑ Ⓒ Ⓓ Ⓔ

6. If $\dfrac{5}{x + 3} = \dfrac{1}{x} + \dfrac{1}{2x}$, what is the value of x ?

    (A) $\dfrac{3}{14}$ (B) $\dfrac{1}{3}$ (C) $\dfrac{6}{13}$ (D) $\dfrac{3}{4}$ (E) $\dfrac{9}{7}$

    Ⓐ Ⓑ Ⓒ Ⓓ Ⓔ

7. If $8^x = 16^{x-1}$, then $x =$

    (A) $\dfrac{1}{8}$ (B) $\dfrac{1}{2}$ (C) 2   (D) 4   (E) 8

    Ⓐ Ⓑ Ⓒ Ⓓ Ⓔ

8. If $a = \dfrac{b+x}{c+x}$, what is the value of x in terms of a, b, and c ?

   (A) $\dfrac{a-bc}{a-1}$

   (B) $\dfrac{b-ac}{a-1}$

   (C) $\dfrac{a+bc}{a+1}$

   (D) $\dfrac{ac+b}{a+1}$

   (E) $\dfrac{ac-b}{a}$ Ⓐ Ⓑ Ⓒ Ⓓ Ⓔ

9. If $2x - 9y = 11$ and $x + 12y = -8$, what is the value of $x + y$?

   (A) $-\dfrac{29}{11}$ (B) $-\dfrac{9}{11}$ (C) 1 (D) $\dfrac{20}{11}$ (E) $\dfrac{29}{11}$

   Ⓐ Ⓑ Ⓒ Ⓓ Ⓔ

10. Which of the following is the solution set of $|2x - 3| < 7$?

   (A) $\{x: -5 < x < 2\}$
   (B) $\{x: -5 < x < 5\}$
   (C) $\{x: -2 < x < 5\}$
   (D) $\{x: x < -5 \text{ or } x > 2\}$
   (E) $\{x: x < -2 \text{ or } x > 5\}$ Ⓐ Ⓑ Ⓒ Ⓓ Ⓔ

Find Your Study Plan

The answer key shows where in this chapter to find explanations for the questions you missed. Here's how you should proceed based on your Diagnostic Test score.

9–10: Superb! You really know your algebra. Unless you have lots of time and just love to read about algebra, you might consider skipping this chapter. You seem to know it all already. To make absolutely sure, you could look over the facts, formulas, and strategies in the margins of this chapter. And if you just want to do some more algebra problems, go to the Follow-Up Test at the end of this chapter.

7–8: Excellent! You're quite good at algebra. Some of these are especially difficult questions. If you're taking a "shortcut" or you're on the Panic Plan, you don't really have time to study this chapter, and you don't really need to. You might want to look at those pages that address the questions you didn't get right. And if you just want to do some more algebra problems, go to the Follow-Up Test at the end of the chapter.

0–6: No matter how pressed for time you are, you should continue to read this chapter and do the Follow-Up Test at the end. You need to brush up on your algebra before you can take full advantage of later chapters.

ALGEBRA TEST TOPICS

The questions in the Algebra Diagnostic Test are typical of those on the Mathematics subject tests. They cover a wide range of algebra topics, from simple expressions to quadratic equations to inequalities with absolute value signs. In this chapter we'll use these questions to review the algebra you need to know. We will also use these questions to demonstrate effective problem-solving techniques, alternative methods, and test-taking strategies that apply to algebra questions.

Evaluating Expressions

The simplest algebra questions on the Mathematics subject tests are like Example 1:

Example 1

If $x = y^2$ and $y = \dfrac{5}{k}$, what is the value of x when $k = \dfrac{1}{2}$?

(A) 2.50   (B) 3.16   (C) 6.25   (D) 20.00   (E) 100.00

Answering a question like this is just a matter of plugging in and cranking out. You are given an expression for x in terms of y, an expression for y in terms of k, and the value of k.

Plug $k = \frac{1}{2}$ into the second equation to find y:

$$y = \frac{5}{k} = \frac{5}{\frac{1}{2}} = 5 \div \frac{1}{2} = \frac{5}{1} \times \frac{2}{1} = 10$$

Now plug $y = 10$ into the first equation to find x:

$$x = y^2 = 10^2 = 100$$

The answer is (E).

Exponents—Key Operations

You can't be adept at algebra unless you're completely at ease with exponents. Here's what you need to know.

Multiplying powers with the same base: To multiply powers with the same base, keep the base and add the exponents:

$$x^3 \times x^4 = x^{3+4} = x^7$$

Dividing powers with the same base: To divide powers with the same base, keep the base and subtract the exponents:

$$y^{13} \div y^8 = y^{13-8} = y^5$$

Raising a power to an exponent: To raise a power to an exponent, keep the base and multiply the exponents:

$$(x^3)^4 = x^{3\times4} = x^{12}$$

Multiplying powers with the same exponent: To multiply powers with the same exponent, multiply the bases and keep the exponent:

$$(3^x)(4^x) = 12^x$$

Dividing powers with the same exponent: To divide powers with the same exponent, divide the bases and keep the exponent:

$$\frac{6^x}{2^x} = 3^x$$

On test day, you might encounter an algebra question, like Example 2, that specifically tests your understanding of the rules of exponents.

> **THE FIVE RULES OF EXPONENTS**
>
> 1. $(x^m)(x^n) = x^{m+n}$
> 2. $\frac{x^m}{x^n} = x^{m-n}$
> 3. $(x^m)^n = x^{mn}$
> 4. $(x^n)(y^n) = (xy)^n$
> 5. $\frac{x^n}{y^n} = \left(\frac{x}{y}\right)^n$

Example 2

For all $xyz \neq 0$, $\dfrac{6x^2 y^{12} z^6}{(2x^2 yz)^3} =$

(A) $\dfrac{y^4 z^2}{x^3}$ (B) $\dfrac{y^9 z^4}{x^4}$ (C) $\dfrac{y^9 z^3}{2x^3}$ (D) $\dfrac{3y^4 z^2}{4x^3}$ (E) $\dfrac{3y^9 z^3}{4x^3}$

There's nothing tricky about this question if you know how to work with exponents. The first step is to eliminate the parentheses. Everything inside gets cubed:

$$\frac{6x^2 y^{12} z^6}{(2x^2 yz)^3} = \frac{6x^2 y^{12} z^6}{8x^6 y^3 z^3}$$

The next step is to look for factors common to the numerator and denominator. The 6 on top and the 8 on bottom reduce to 3 over 4—so it already looks like the answer's going to be (D) or (E). The x^2 on top cancels with the x^6 on bottom, leaving x^4 on bottom.

You're actually subtracting the exponents: $2 - 6 = -4$, since x^{-4} is the same as $\dfrac{1}{x^4}$. The y^{12} on top cancels with the y^3 on bottom, leaving y^9 on top. And the z^6 on top cancels with the z^3 on bottom, leaving z^3 on top:

$$\frac{6x^2 y^{12} z^6}{8x^6 y^3 z^3} = \frac{3y^9 z^3}{4x^4}$$

The answer is (E).

Adding, Subtracting, and Multiplying Polynomials

Algebra is the basic language of the Mathematics Subject Tests, and you will want to be fluent in that language. You might not get a whole lot of questions that ask explicitly about such basic algebra procedures as combining like terms, multiplying binomials, or factoring algebraic expressions, but you will do all of those things in the course of working out the answers to more advanced questions. So it's essential that you be at ease with the mechanics of algebraic manipulations.

Combining like terms: To combine like terms, keep the variable part unchanged while adding or subtracting the coefficients:

$$2a + 3a = (2 + 3)a = 5a$$

Adding or subtracting polynomials: To add or subtract polynomials, combine like terms:

$$(3x^2 + 5x - 7) - (x^2 + 12) =$$
$$(3x^2 - x^2) + 5x + (-7 - 12) =$$
$$2x^2 + 5x - 19$$

> **COMBINING LIKE TERMS**
>
> $ax + bx = (a + b)x$
> $ax - bx = (a - b)x$

Multiplying monomials: To multiply monomials, multiply the coefficients and the variables separately:

$$2x \times 3x = (2 \times 3)(x \times x) = 6x^2$$

Multiplying binomials: To multiply binomials, use FOIL. To multiply $(x + 3)$ by $(x + 4)$, first multiply the **F**irst terms: $x \times x = x^2$. Next the **O**uter terms: $x \times 4 = 4x$. Then the **I**nner terms: $3 \times x = 3x$. And finally the **L**ast terms: $3 \times 4 = 12$. Then add and combine like terms:

$$(x + 3)(x + 4) = x^2 + 4x + 3x + 12 = x^2 + 7x + 12$$

Multiplying polynomials: To multiply polynomials with more than two terms, make sure you multiply each term in the first polynomial by each term in the second. (FOIL works only when you want to multiply two binomials.)

$$
\begin{aligned}
(x^2 + 3x + 4)(x + 5) &= x^2(x + 5) + 3x(x + 5) + 4(x + 5) \\
&= x^3 + 5x^2 + 3x^2 + 15x + 4x + 20 \\
&= x^3 + 8x^2 + 19x + 20
\end{aligned}
$$

After multiplying two polynomials together, the number of terms in your expression before simplifying should equal the number of terms in one polynomial multiplied by the number of terms in the second. In the example above, you should have $3 \times 2 = 6$ terms in the product before you simplify like terms.

Dividing Polynomials

To divide polynomials, you can use long division. For example, to divide $2x^3 + 13x^2 + 11x - 16$ by $x + 5$:

$$x + 5 \overline{)2x^3 + 13x^2 + 11x - 16}$$

The first term of the quotient is $2x^2$, because that's what will give you a $2x^3$ as a first term when you multiply it by $x + 5$:

$$
\begin{array}{r}
2x^2 \\
x + 5 \overline{)2x^3 + 13x^2 + 11x - 16} \\
\underline{2x^3 + 10x^2}
\end{array}
$$

MULTIPLYING MONOMIALS

$(ax)(bx) = (ab)x^2$

MULTIPLYING BINOMIALS—FOIL

$(a + b)(c + d) = ?$

First $= ac$

Outer $= ad$

Inner $= bc$

Last $= bd$

Product $= ac + ad + bc + bd$

Subtract and continue in the same way as when dividing numbers:

$$
\begin{array}{r}
2x^2 + 3x - 4 \\
x+5 \overline{\smash{\big)}\ 2x^3 + 13x^2 + 11x - 16} \\
\underline{2x^3 + 10x^2} \\
3x^2 + 11x \\
\underline{3x^2 + 15x} \\
-4x - 16 \\
\underline{-4x - 20} \\
4
\end{array}
$$

The result is $2x^2 + 3x - 4$ with a remainder of 4.

Long division is the way to do Example 3:

Example 3

When $2x^3 + 3x^2 - 4x + k$ is divided by $x + 2$, the remainder is 3. What is the value of k ?

(A) –1   (B) 1   (C) 2   (D) 3   (E) 5

To answer this question, start by cranking out the long division:

$$
\begin{array}{r}
2x^2 - x - 2 \\
x+2 \overline{\smash{\big)}\ 2x^3 + 3x^2 - 4x + k} \\
\underline{2x^3 + 4x^2} \\
-x^2 - 4x \\
\underline{-x^2 - 2x} \\
-2x + k \\
\underline{-2x - 4}
\end{array}
$$

The question says that the remainder is 3, so whatever k is, when you subtract –4 from it, you get 3:

$$
k - (-4) = 3
$$
$$
k + 4 = 3
$$
$$
k = -1
$$

The answer is (A).

Factoring

Performing operations on polynomials is largely a matter of cranking it out. Once you know the rules, adding, subtracting, multiplying, and even dividing is automatic. Factoring algebraic expressions is a different matter.

To factor successfully, you have to do more thinking and less cranking. You have to try to figure out what expressions multiplied will give you the polynomial you're looking at. Sometimes that means having a good eye for the test makers' favorite factorables:

- Factor common to all terms
- Difference of squares
- Square of a binomial

Factor common to all terms: A factor common to all the terms of a polynomial can be factored out. This is essentially the distributive property in reverse. For example, all three terms in the polynomial $3x^3 + 12x^2 - 6x$ contain a factor of $3x$. Pulling out the common factor yields $3x(x^2 + 4x - 2)$.

Difference of squares: You will want to be especially keen at spotting polynomials in the form of the difference of squares. Whenever you have two identifiable squares with a minus sign between them, you can factor the expression like this:

$$a^2 - b^2 = (a + b)(a - b)$$

For example, $4x^2 - 9$ factors to $(2x + 3)(2x - 3)$.

Squares of binomials: Learn to recognize polynomials that are squares of binomials:

$$a^2 + 2ab + b^2 = (a + b)^2$$
$$a^2 - 2ab + b^2 = (a - b)^2$$

For example, $4x^2 + 12x + 9$ factors to $(2x + 3)^2$, and $a^2 - 10a + 25$ factors to $(a - 5)^2$.

Sometimes you'll want to factor a polynomial that's not in any of these classic factorable forms. When that happens, factoring becomes a kind of logic exercise, with some trial and error thrown in. To factor a quadratic expression, think about what binomials you could use FOIL on to get that quadratic expression. For example, to factor $x^2 - 5x + 6$, think about what **F**irst terms will produce x^2, what **L**ast terms will produce $+6$, and what **O**uter and **I**nner terms will produce $-5x$. Some common sense—and a little trial and error—will lead you to $(x - 2)(x - 3)$.

Example 4 is a good instance of a Math 1 question that calls for factoring.

> ## CLASSIC FACTORABLES
>
> Factor common to all terms:
> $ax + ay = a(x + y)$
> Difference of squares:
> $a^2 - b^2 = (a - b)(a + b)$
> Square of binomial:
> $a^2 + 2ab + b^2 = (a + b)^2$

Example 4

For all $x \neq \pm 3$, $\dfrac{3x^2 - 11x + 6}{9 - x^2} =$

(A) $\dfrac{2 - 3x}{x - 3}$ (B) $\dfrac{2 - 3x}{x + 3}$ (C) $\dfrac{2x - 3}{x - 3}$ (D) $\dfrac{3x - 2}{x + 3}$ (E) $\dfrac{3x - 2}{x - 3}$

To reduce a fraction, you eliminate factors common to the top and bottom. So the first step in reducing an algebraic fraction is to *factor the numerator and denominator*. Here the denominator is easy since it's the difference of squares: $9 - x^2 = (3 - x)(3 + x)$. The numerator takes some thought and some trial and error. For the first term to be $3x^2$, the first terms of the factors must be $3x$ and x. For the last term to be $+6$, the last terms must be either $+2$ and $+3$, or -2 and -3, or $+1$ and $+6$, or -1 and -6. After a few tries, you should come up with: $3x^2 - 11x + 6 = (3x - 2)(x - 3)$. Now the fraction looks like this:

$$\frac{3x - 11x + 6}{9 - x^2} = \frac{(3x - 2)(x - 3)}{(3 - x)(3 + x)}$$

In this form there are no precisely common factors, but there is a factor in the numerator that's the opposite (negative) of a factor in the denominator: $x - 3$ and $3 - x$ are opposites. Factor -1 out of the numerator and get:

$$\frac{(3x - 2)(x - 3)}{(3 - x)(3 + x)} = \frac{(-1)(3x - 2)(3 - x)}{(3 - x)(3 + x)}$$

Now $(3 - x)$ can be eliminated from both the top and the bottom:

$$\frac{(-1)(3x - 2)(3 - x)}{(3 - x)(3 + x)} = \frac{-(3x - 2)}{3 + x} = \frac{-3x + 2}{3 + x}$$

That's the same as choice (B):

$$\frac{-3x + 2}{3 + x} = \frac{2 - 3x}{x + 3}$$

Alternative method: Here's another way to answer this question. *Pick a number for x and see what happens*. One of the answer choices will give you the same value as the original fraction will, no matter what you plug in for *x*. Pick a number that's easy to work with—like 0.

When you plug $x = 0$ into the original expression, any term with an x drops out, and you end up with $\dfrac{6}{9}$, or $\dfrac{2}{3}$. Now plug $x = 0$ into each answer choice to see which ones equal $\dfrac{2}{3}$.

When you get to (B), it works, but you can't stop there. It might just be a coincidence. When you pick numbers, *look at every answer choice*. Choice (E) also works for $x = 0$. At least you know one of those is the correct answer, and you can decide between them by picking another value for *x*.

This is not a sophisticated approach, but who cares? You don't get points for elegance. You get points for right answers.

The Golden Rule of Equations

You probably remember the basic procedure for solving algebraic equations: *Do the same thing to both sides.* You can do almost anything you want to one side of an equation as long as you preserve the equality by doing the same thing to the other side. Your aim in whatever you do to both sides is to get the variable (or expression) you're solving for all by itself on one side. Look at Example 5.

Example 5

If $\sqrt[3]{8x+6} = -3$, what is the value of x ?

(A) -4.125 (B) -2.625 (C) -1.875 (D) -1.125 (E) 2.625

To solve this equation for x, you do whatever necessary to both sides of the equation to get x all by itself on one side. Layer by layer you want to peel away all those extra symbols and numbers around the x. First you want to get rid of that cube-root symbol. The way to undo a cube root is to cube both sides:

$$\sqrt[3]{8x+6} = -3$$
$$\left(\sqrt[3]{8x+6}\right)^3 = (-3)^3$$
$$8x+6 = -27$$

The rest is easy. Subtract 6 from both sides and divide both sides by 8:

$$8x+6 = -27$$
$$8x = -27-6$$
$$8x = -33$$
$$x = -\frac{33}{8} = -4.125$$

The answer is (A).

The test makers have a couple of favorite equation types that you should be prepared to solve. Solving linear equations is usually pretty straightforward. Generally it's obvious what to do to isolate the unknown. But when the unknown is in a denominator or an exponent, it might not be so obvious how to proceed.

> ## PICK NUMBERS
>
> When the answer choices are algebraic expressions, it is often quickest to *pick numbers* for the unknowns, plug those numbers into the stem and see what you get, and then plug those same numbers into the answer choices to find matches.
>
> *Warning:* When you pick numbers, you have to check *all* the answer choices. Sometimes more than one works with the number(s) you pick, in which case you have to pick numbers again.

Unknown in a Denominator

The basic procedure for solving an equation is the same even when the unknown is in a denominator: Do the same thing to both sides. In this case you multiply in order to undo division.

If you wanted to solve $1 + \dfrac{1}{x} = 2 - \dfrac{1}{x}$, you would multiply both sides by x:

$$1 + \frac{1}{x} = 2 - \frac{1}{x}$$

$$x\left(1 + \frac{1}{x}\right) = x\left(2 - \frac{1}{x}\right)$$

$$x + 1 = 2x - 1$$

Now you have an equation with no denominators, which is easy to solve:

$$x + 1 = 2x - 1$$
$$x - 2x = -1 - 1$$
$$-x = -2$$
$$x = 2$$

Another good way to solve an equation with the unknown in the denominator is to *cross multiply*. That's the best way to do Example 6.

Example 6

If $\dfrac{5}{x+3} = \dfrac{1}{x} + \dfrac{1}{2x}$, what is the value of x?

(A) $\dfrac{3}{14}$ (B) $\dfrac{1}{3}$ (C) $\dfrac{6}{13}$ (D) $\dfrac{3}{4}$ (E) $\dfrac{9}{7}$

Before you can cross multiply, you need to express the right side of the equation as a single fraction. That means giving the two fractions a common denominator and adding them. The common denominator is $2x$:

$$\frac{5}{x+3} = \frac{1}{x} + \frac{1}{2x}$$

$$\frac{5}{x+3} = \frac{2}{2x} + \frac{1}{2x}$$

$$\frac{5}{x+3} = \frac{3}{2x}$$

Now you can cross multiply:

$$\frac{5}{x+3}=\frac{3}{2x}$$

$$(5)(2x)=(x+3)(3)$$

$$10x=3x+9$$

$$10x-3x=9$$

$$7x=9$$

$$x=\frac{9}{7}$$

The answer is (E).

Unknown in an Exponent

The procedure for solving an equation when the unknown is in an exponent is a little different. What you want to do in this situation is to express one or both sides of the equation so that the two sides have the same base. Look at Example 7.

Example 7

If $8^x = 16^{x-1}$, then $x =$

(A) $\frac{1}{8}$ (B) $\frac{1}{2}$ (C) 2   (D) 4   (E) 8

In this case, the base on the left is 8 and the base on the right is 16. They're both powers of 2, so you can express both sides as powers of 2:

$$8^x=16^{x-1}$$

$$(2^3)^x=(2^4)^{x-1}$$

$$2^{3x}=2^{4x-4}$$

Now that both sides have the same base, you can simply set the exponent expressions equal and solve for x:

$$3x=4x-4$$

$$3x-4x=-4$$

$$-x=-4$$

$$x=4$$

The answer is (D).

Alternative method: This is a good place to Backsolve. Try plugging the answer choices back into the problem until you find the one that works. In this case, work with the easier, whole numbers. If you start with (C) and $x = 2$, you get $8^x = 8^2 = 64$ on the left side of the equation and $16^{x-1} = 16^1 = 16$ on the right side. It's not clear whether (C) was too small or too large, so you should probably try (D) next—it's easier to work with than (B), which is a fraction. If $x = 4$, then $8^x = 8^4 = 4{,}096$ on the left, and $16^{x-1} = 16^3 = 4{,}096$. No need to do any more. (D) works, so it's the answer.

Don't depend on backsolving too much. There are lots of math questions that can't be backsolved at all. And most that *can* be backsolved are almost certainly more *quickly* solved by a more direct approach.

Quadratic Equations

To solve a quadratic equation, put it in the "$ax^2 + bx + c = 0$" form, factor the left side (if you can), and set each factor equal to 0 separately to get the two solutions. To solve $x^2 + 12 = 7x$, first rewrite it as $x^2 - 7x + 12 = 0$. Then factor the left side:

$$x^2 - 7x + 12 = 0$$
$$(x - 3)(x - 4) = 0$$
$$x - 3 = 0 \text{ or } x - 4 = 0$$
$$x = 3 \text{ or } 4$$

Sometimes the left side may not be obviously factorable. You can always use the *quadratic formula*. Just plug the coefficients a, b, and c from $ax^2 + bx + c = 0$ into the formula:

$$x = \frac{-b \pm \sqrt{b^2 - 4ac}}{2a}$$

To solve $x^2 + 4x + 2 = 0$, plug $a = 1$, $b = 4$, and $c = 2$ into the formula:

$$x = \frac{-4 \pm \sqrt{4^2 - 4 \cdot 1 \cdot 2}}{2 \cdot 1}$$

$$= \frac{-4 \pm \sqrt{8}}{2} = -2 \pm \sqrt{2}$$

"In Terms Of"

So far in this chapter, solving an equation has meant finding a numerical value for the unknown. When there's more than one variable, it's generally impossible to get numerical solutions. Instead, what you do is solve for the unknown *in terms of* the other variables.

To solve an equation for one variable in terms of another means to isolate the one variable on one side of the equation, leaving an expression containing the other variable on the other side of the equation.

For example, to solve the equation $3x - 10y = -5x + 6y$ for x in terms of y, isolate x:

$$3x - 10y = -5x + 6y$$
$$3x + 5x = 6y + 10y$$
$$8x = 16y$$
$$x = 2y$$

Now look at the next example, which asks you to solve "in terms of."

Example 8

If $a = \dfrac{b+x}{c+x}$, what is the value of x in terms of a, b, and c?

(A) $\dfrac{a-bc}{a-1}$ (B) $\dfrac{b-ac}{a-1}$ (C) $\dfrac{a+bc}{a+1}$ (D) $\dfrac{ac+b}{a+1}$ (E) $\dfrac{ac-b}{a}$

You want to get x on one side by itself. First thing to do is eliminate the denominator by multiplying both sides by $c + x$:

$$a = \frac{b+x}{c+x}$$
$$a(c+x) = \left(\frac{b+x}{c+x}\right)(c+x)$$
$$ac + ax = b + x$$

Next move all terms with x to one side and all terms without x to the other:

$$ac + ax = b + x$$
$$ax - x = b - ac$$

Now factor x out of the left side and divide both sides by the other factor to isolate x:

$$ax - x = b - ac$$
$$x(a-1) = b - ac$$
$$x = \frac{b-ac}{a-1}$$

The answer is (B).

Simultaneous Equations

You can get numerical solutions for more than one unknown if you are given the same number of equations as unknowns. The test makers like simultaneous equations questions because they take a little thought to answer. Solving simultaneous equations almost always involves combining equations, but you have to figure out what's the best way to combine the equations.

> **DON'T DO MORE WORK THAN YOU HAVE TO**
>
> You don't always have to find the value of each variable to answer a simultaneous equations question.

You can solve for two variables only if you have two distinct equations. Two forms of the same equation will not be adequate. Combine the equations in such a way that one of the variables cancels out. For example, to solve the two equations $4x + 3y = 8$ and $x + y = 3$, multiply both sides of the second equation by -3 to get $-3x - 3y = -9$. Now add the two equations; the $3y$ and the $-3y$ cancel out, leaving $x = -1$. Plug that back into either one of the original equations, and you'll find that $y = 4$.

Example 9 is a simultaneous equations question.

Example 9

If $2x - 9y = 11$ and $x + 12y = -8$, what is the value of $x + y$?

(A) $-\dfrac{29}{11}$ (B) $-\dfrac{9}{11}$ (C) 1   (D) $\dfrac{20}{11}$ (E) $\dfrac{29}{11}$

If you just plow ahead without thinking, you might try to answer this question by solving for one variable at a time. That would work, but it would take a lot more time than this question needs. As usual, the key to this simultaneous equations question is to combine the equations, but combining the equations doesn't necessarily mean losing a variable. Look what happens here if you just add the equations as presented:

$$2x - 9y = 11$$
$$+[x + 12y = -8]$$
$$\overline{3x + 3y = 3}$$

Suddenly you're almost there! Just divide both sides by 3 and you get $x + y = 1$. The answer is (C).

Absolute Value

To solve an equation that includes absolute value signs, think about the two different cases. For example, to solve the equation $|x - 12| = 3$, think of it as two equations:

$$x - 12 = 3 \text{ or } x - 12 = -3$$
$$x = 15 \text{ or } 9$$

Inequalities

To solve an inequality, do whatever is necessary to both sides to isolate the variable. Just remember that when you multiply or divide both sides by a negative number, you must reverse the sign. To solve $-5x + 7 < -3$, subtract 7 from both sides to get $-5x < -10$. Now divide both sides by -5, remembering to reverse the sign: $x > 2$.

Inequalities and Absolute Value

About the most complicated algebraic solving you'll have to do on the Math 1 test will involve inequalities and absolute value signs. Look at Example 10.

> **INEQUALITIES AND ABSOLUTE VALUE**
>
> For all $n > 0$,
> if $|x| < n$, then
> $-n < x < n$;
> if $|x| > n$, then
> $x < -n$ or $x > n$.

Example 10

Which of the following is the solution set of $|2x - 3| < 7$?

(A) $\{x: -5 < x < 2\}$
(B) $\{x: -5 < x < 5\}$
(C) $\{x: -2 < x < 5\}$
(D) $\{x: x < -5 \text{ or } x > 2\}$
(E) $\{x: x < -2 \text{ or } x > 5\}$

What does it mean if $|2x - 3| < 7$? It means that if the expression between the absolute value bars is positive, it's less than $+7$ or, if the expression between the bars is negative, it's greater than -7. In other words, $2x - 3$ is between -7 and $+7$:

$$-7 < 2x - 3 < 7$$
$$-4 < 2x < 10$$
$$-2 < x < 5$$

The answer is (C).

In fact, there's a general rule that applies here: To solve an inequality in the form $|\text{whatever}| < p$, where $p > 0$, just put that "whatever" inside the range $-p$ to p:

$$|\text{whatever}| < p \text{ means: } -p < \text{whatever} < p$$

For example, $|x - 5| < 14$ becomes $-14 < x - 5 < 14$.

And here's another general rule: To solve an inequality in the form $|\text{whatever}| > p$, where $p > 0$, just put that "whatever" outside the range $-p$ to p:

$$|\text{whatever}| > p \text{ means: whatever} < -p \text{ OR whatever} > p$$

For example, $\left|\dfrac{3x+9}{2}\right| > 7$ becomes $\dfrac{3x+9}{2} < -7$ OR $\dfrac{3x+9}{2} > 7$.

Well, you've seen a lot of algebra in this chapter. You've seen ten of the test makers' favorite algebra situations. You've reviewed all the relevant algebra facts and formulas. And you've learned some effective Kaplan test-taking strategies. Now it's time to take the Algebra Follow-Up Test to find out how much you've learned.

THINGS TO REMEMBER:

The Rules of Exponents

1. $(x^m)(x^n) = x^{m+n}$

2. $\dfrac{x^m}{x^n} = x^{m-n}$

3. $(x^m)^n = x^{mn}$

4. $(x^n)(y^n) = (xy)^n$

5. $\dfrac{x^n}{y^n} = \left(\dfrac{x}{y}\right)^n$

Combining Like Terms

$ax + bx = (a + b)x$
$ax - bx = (a - b)x$

Multiplying Monomials

$(ax)(bx) = (ab)x^2$

Multiplying Binomials—FOIL

$(a + b)(c + d) = ?$
First = ac
Outer = ad
Inner = bc
Last = bd
Product = $ac + ad + bc + bd$

Classic Factorables

Factor common to all terms:
$$ax + ay = a(x + y)$$

Difference of squares:
$$a^2 - b^2 = (a - b)(a + b)$$

Square of binomial:
$$a^2 + 2ab + b^2 = (a + b)^2$$

Quadratic Formula

If $ax^2 + bx + c = 0$, then:
$$x = \frac{-b \pm \sqrt{b^2 - 4ac}}{2a}$$

Inequalities and Absolute Value

For all $n > 0$,
  if $|x| < n$, then
  $-n < x < n$;
  if $|x| > n$, then
  $x < -n$ or $x > n$.

ALGEBRA FOLLOW-UP TEST

10 Questions (12 Minutes)

Directions: Solve the following problems. Fill in the oval corresponding to the best answer choice in the grid to the right of each question. (Answers and explanations begin on page 58.)

DO YOUR FIGURING HERE

1. If $x = 3 - y^2$ and $y = -2$, what is the value of x ?

    (A) -2 (B) -1 (C) 1 (D) 2 (E) 25

    Ⓐ Ⓑ Ⓒ Ⓓ Ⓔ

2. For all x, $2^x + 2^x + 2^x + 2^x =$

    (A) 2^{x+2} (B) 2^{x+4} (C) 2^{3x} (D) 2^{4x} (E) 2^{5x}

    Ⓐ Ⓑ Ⓒ Ⓓ Ⓔ

3. For all $x \neq \pm\dfrac{1}{2}, \dfrac{6x^2 - x - 2}{4x^2 - 1} =$

    (A) $\dfrac{2 - 3x}{2x + 1}$

    (B) $\dfrac{3x + 2}{2x + 1}$

    (C) $\dfrac{3x + 2}{2x - 1}$

    (D) $\dfrac{3x - 2}{2x + 1}$

    (E) $\dfrac{3x - 2}{2x - 1}$

    Ⓐ Ⓑ Ⓒ Ⓓ Ⓔ

4. When $3x^3 - 7x + 7$ is divided by $x + 2$, the remainder is

    (A) -5 (B) -3 (C) 1 (D) 3 (E) 5

    Ⓐ Ⓑ Ⓒ Ⓓ Ⓔ

KAPLAN

5. If $\sqrt[4]{\dfrac{x+1}{2}} = \dfrac{1}{2}$, then $x =$

    (A) −0.969   (B) −0.875   (C) 0   (D) 0.875   (E) 0.969

    Ⓐ Ⓑ Ⓒ Ⓓ Ⓔ

6. If $\dfrac{19}{5x+17} = \dfrac{19}{31}$, then $x =$

    (A) 0.4   (B) 1.4   (C) 2.8   (D) 3.4   (E) 3.8

    Ⓐ Ⓑ Ⓒ Ⓓ Ⓔ

7. If $(3^{x^2})(9^x)(3) = 27$ and $x > 0$, what is the value of x ?

    (A) 0.268   (B) 0.414   (C) 0.732   (D) 1.414   (E) 1.464

    Ⓐ Ⓑ Ⓒ Ⓓ Ⓔ

8. If $y \neq 4a$, and $x = \dfrac{y+a^2}{y-4a}$, what is the value of y in terms of a and x ?

    (A) $\dfrac{4a - 4a^2x}{x+1}$

    (B) $\dfrac{a^2 - 4ax}{x+1}$

    (C) $\dfrac{a^2 + 4ax}{x+1}$

    (D) $\dfrac{a^2 + 4ax}{x-1}$

    (E) $\dfrac{a^2 - 4ax}{x-1}$ Ⓐ Ⓑ Ⓒ Ⓓ Ⓔ

9. If one of the following choices is the solution to the pair of equations $4x + ky = 15$ and $x - ky = -25$, which one is it?

    (A) $x = -3$ and $y = -5$
    (B) $x = -2$ and $y = 3$
    (C) $x = 0$ and $y = -2$
    (D) $x = 2$ and $y = 3$
    (E) $x = 3$ and $y = 5$ Ⓐ Ⓑ Ⓒ Ⓓ Ⓔ

10. How many integers are in the solution set of $|4x + 3| < 8$?

    (A)   None
    (B)   Two
    (C)   Three
    (D)   Four
    (E)   Infinitely many   Ⓐ Ⓑ Ⓒ Ⓓ Ⓔ

DO YOUR FIGURING HERE

**Turn the page
for answers and explanations
to the Follow-Up Test.**

FOLLOW-UP TEST—ANSWERS AND EXPLANATIONS

1. B

Plug $y = -2$ into the first equation:

$$x = 3 - y^2 = 3 - (-2)^2 = 3 - 4 = -1$$

2. A

The sum of four identical quantities is 4 times one of those quantities, so the sum of the four terms 2^x is 4 times 2^x:

$$2^x + 2^x + 2^x + 2^x = 4(2^x) = 2^2(2^x) = 2^{x+2}$$

3. E

Factor the top and the bottom and cancel the factors they have in common:

$$\frac{6x^2 - x - 2}{4x^2 - 1} = \frac{(3x-2)(2x+1)}{(2x-1)(2x+1)} = \frac{3x-2}{2x-1}$$

4. B

Use long division. Watch out: The expression that goes under the division sign needs a place-holding $0x^2$ term:

$$
\begin{array}{r}
3x^2 - 6x + 5 \\
x+2 \overline{\smash{\big)}\, 3x^3 + 0x^2 - 7x + 7} \\
\underline{3x^3 + 6x^2} \\
-6x^2 - 7x \\
\underline{-6x^2 - 12x} \\
5x + 7 \\
\underline{5x + 10} \\
-3
\end{array}
$$

The remainder is –3.

5. B

To undo the fourth-root symbol, raise both sides to the fourth power:

$$\sqrt[4]{\frac{x+1}{2}} = \frac{1}{2}$$

$$\left(\sqrt[4]{\frac{x+1}{2}}\right)^4 = \left(\frac{1}{2}\right)^4$$

$$\frac{x+1}{2} = \frac{1}{16}$$

Now cross multiply:

$$\frac{x+1}{2} = \frac{1}{16}$$
$$(x+1)(16) = (2)(1)$$
$$16x + 16 = 2$$
$$16x = -14$$
$$x = -\frac{14}{16} = -\frac{7}{8} = -0.875$$

6. C

This might look at first glance like a candidate for cross multiplication, but that would just make things more complicated. Notice that the fractions on both sides have the same numerator, 19. So the numerator is irrelevant. If the two fractions are equal and they have the same numerator, then they must have the same denominator. So just write an equation that says that one denominator is equal to the other denominator:

$$\frac{19}{5x+17} = \frac{19}{31}$$
$$5x + 17 = 31$$
$$5x = 31 - 17$$
$$5x = 14$$
$$x = \frac{14}{5} = 2.8$$

7. C

Watch what happens when you express everything as powers of 3:

$$3^{x^2}(9^x)3 = 27$$
$$3^{x^2}(3^{2x})3^1 = 3^3$$

The left side of the equation is the product of powers with the same base, so just add the exponents:

$$3^{x^2}(3^{2x})3^1 = 3^3$$
$$3^{x^2+2x+1} = 3^3$$

Now the two sides of the equation are powers with the same base, so you can just set the exponents equal:

$$3^{x^2+2x+1} = 3^3$$
$$x^2 + 2x + 1 = 3$$
$$(x+1)^2 = 3$$
$$x + 1 = \pm\sqrt{3}$$

The positive value is $\sqrt{3} - 1$, which is approximately 0.732.

8. D

First multiply both sides by $y - 4a$ to clear the denominator:

$$x = \frac{y+a^2}{y-4a}$$
$$x(y-4a) = y+a^2$$
$$xy - 4ax = y + a^2$$

Now move all terms with y to the left and all terms without y to the right:

$$xy - 4ax = y + a^2$$
$$xy - y = a^2 + 4ax$$

Now factor the left side and divide to isolate y :

$$xy - y = a^2 + 4ax$$
$$y(x-1) = a^2 + 4ax$$
$$y = \frac{a^2+4ax}{x-1}$$

9. B

With only two equations, you won't be able to get numerical solutions for three unknowns. But apparently you can get far enough to rule out four of the five answer choices. How? Look for a way to combine the equations that leads somewhere useful. Notice that the first equation contains $+ ky$ and the second equation contains $- ky$, so if you add the equations as they are, you'll lose those terms:

$$4x + ky = 15$$
$$x - ky = -25$$
$$5x = -10$$
$$x = -2$$

There's not enough information to get numerical solutions for k or y, but you do know that $x = -2$, so the correct answer is the only choice that has an x-coordinate of -2.

10. D

If the absolute value of something is less than 8, then that something is between –8 and 8:

$$|4x + 3| < 8$$
$$-8 < 4x + 3 < 8$$
$$-11 < 4x < 5$$
$$-\frac{11}{4} < x < \frac{5}{4}$$
$$-2\frac{3}{4} < x < 1\frac{1}{4}$$

There are four integers in that range: –2, –1, 0, and 1.

Chapter 5: **Plane Geometry**

- Line segments
- Triangles
- Quadrilaterals and polygons
- Circles

According to the official breakdown, almost 20 percent of Math 1 questions are plane geometry questions. But that's counting only the ones that are explicitly and primarily plane geometry questions. In fact, plane geometry is fundamental to solid geometry, coordinate geometry, and trigonometry. The material in this chapter is relevant to nearly half of the Math 1 test.

> **PLANE GEOMETRY FACTS AND FORMULAS IN THIS CHAPTER**
>
> - Five Facts about Triangles
> - Similar Figures
> - Three Special Triangle Types
> - Pythagorean Theorem
> - Four Special Right Triangles
> - Five Special Quadrilateral Types
> - Polygon Angles
> - Four Circle Formulas

HOW TO USE THIS CHAPTER

Maybe you already know all the plane geometry you need. You can find out by taking the Plane Geometry Diagnostic Test. The six plane geometry questions on the Diagnostic Test are typical of those on the Mathematics subject tests. Check your answers using the answer key following the test. No matter how you score, don't worry! The answer key also shows where to find a detailed explanation for each question. The "Find Your Study Plan" section that follows the test will suggest the next steps based on your performance on the Diagnostic.

Find Your Level

How you use this chapter really depends on how much time you have to prep. Find your level and pace below.

Standard Plan. No matter how well you do on the Plane Geometry Diagnostic Test, read the rest of this chapter and do all the practice problems.

Shortcut: Take the Plane Geometry Diagnostic Test and check your answers. If you can answer at least four of the six questions correctly, then you already know the material in this chapter well enough to move on.

Panic Plan. Take the Plane Geometry Diagnostic Test and check your answers. The material in this chapter is vital. Don't try to move on to solid geometry, coordinate geometry, or trigonometry until you feel comfortable with the material in this chapter.

PLANE GEOMETRY DIAGNOSTIC TEST

6 Questions (8 Minutes)

Directions: Solve the following problems. Fill in the oval corresponding to the best answer choice in the grid to the right of each question. (Answers are on page 66.)

DO YOUR FIGURING HERE

1. In Figure 1, the length of $\overline{PS}$ is $2x + 12$, and the length of $\overline{PQ}$ is $6x - 10$. If R is the midpoint of $\overline{QS}$, what is the length of $\overline{PR}$?

   (A) $-4x + 22$
   (B) $-2x + 11$
   (C) $-2x + 22$
   (D) $2x + 22$
   (E) $4x + 1$ Ⓐ Ⓑ Ⓒ Ⓓ Ⓔ

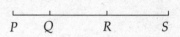

Figure 1

2. In Figure 2, $AB = BC$. If the area of $\triangle ABE$ is x, what is the area of $\triangle ACD$?

   (A) $x\sqrt{2}$
   (B) $x\sqrt{3}$
   (C) $2x$
   (D) $3x$
   (E) $4x$ Ⓐ Ⓑ Ⓒ Ⓓ Ⓔ

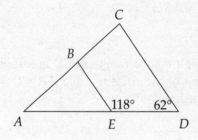

Figure 2

3. In Figure 3, $\triangle ABC$ is equilateral and $\triangle ADC$ is isosceles. If $AC = 1$, what is the distance from B to D?

   (A)   0.286
   (B)   0.318
   (C)   0.333
   (D)   0.366
   (E)   0.383   Ⓐ Ⓑ Ⓒ Ⓓ Ⓔ

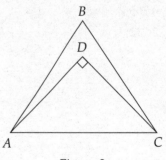

Figure 3

DO YOUR FIGURING HERE

4. In Figure 4, the perimeter of isosceles trapezoid *ABCD* is 50. If *BC* = 9 and *AD* = 21, what is the length of diagonal $\overline{AC}$?

   (A)   13
   (B)   14
   (C)   15
   (D)   16
   (E)   17   Ⓐ Ⓑ Ⓒ Ⓓ Ⓔ

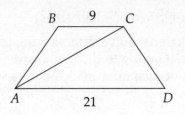

Figure 4

5. A square and a regular hexagon have the same perimeter. If the area of the square is 2.25, what is the area of the hexagon?

   (A)   2.250
   (B)   2.598
   (C)   2.838
   (D)   3.464
   (E)   3.375   Ⓐ Ⓑ Ⓒ Ⓓ Ⓔ

6. In Figure 5, rectangle *ABCD* is inscribed in a circle. If the radius of the circle is 1 and *AB* = 1, what is the area of the shaded region?

   (A)   0.091
   (B)   0.285
   (C)   0.614
   (D)   0.705
   (E)   0.732   Ⓐ Ⓑ Ⓒ Ⓓ Ⓔ

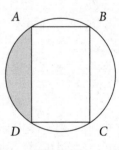

Figure 5

Find Your Study Plan

The answer key on the right shows where in this chapter to find explanations for the questions you missed. Here's how you should proceed based on your Diagnostic Test score.

6: Superb! You really know your plane geometry. Unless you have lots of time and just love to read about plane geometry, you might consider skipping this chapter. You seem to know it all already. Just to make absolutely sure, you could look over the plane geometry facts, formulas, and strategies in the margins of this chapter. And if all you want is more plane geometry questions to try, go to the Follow-Up Test at the end of this chapter.

4–5: Excellent! You're quite good at plane geometry. Some of these are especially difficult questions. If you're taking a "shortcut" or you're on the Panic Plan, you don't really have time to study this chapter, and you don't really need to. You might at least look over those pages that address the questions you didn't get right. You could also look over the plane geometry facts, formulas, and strategies in the margins of this chapter. If you just want to try more plane geometry questions, go to the Follow-Up Test at the end of the chapter.

0–3: You should read this chapter and do the Follow-Up Test at the end. You need to have a good command of the material in this chapter before moving on to later chapters on solid geometry, coordinate geometry, and trigonometry.

DIAGNOSTIC TEST ANSWER KEY

1. E
See "Adding and Subtracting Segment Lengths."

2. E
See "Similar Triangles."

3. D
See "Three Special Triangle Types."

4. E
See "Hidden Special Triangles."

5. B
See "Polygons—Perimeter and Area."

6. C
See "Circles Combined with Other Figures."

PLANE GEOMETRY TEST TOPICS

The questions in the Plane Geometry Diagnostic Test are typical of those on the Math 1 test. They range from segments, to polygons (especially triangles), to circles and multiple figures. In this chapter we'll use these questions to review the plane geometry you're expected to know. We will also use these questions to demonstrate some effective problem-solving techniques, alternative methods, and test-taking strategies that apply to geometry questions.

Adding and Subtracting Segment Lengths

The simplest type of plane geometry questions you may encounter are like Example 1, which involves adding and subtracting segment lengths. It's typical of the Math 1 test that the lengths you have to add and subtract are algebraic expressions rather than numbers.

Example 1

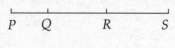

Figure 1

In Figure 1, the length of $\overline{PS}$ is $2x + 12$, and the length of $\overline{PQ}$ is $6x - 10$. If R is the midpoint of $\overline{QS}$, what is the length of $\overline{PR}$?

(A) $-4x + 22$
(B) $-2x + 11$
(C) $-2x + 22$
(D) $2x + 22$
(E) $4x + 1$

Usually the best thing to do to start on a plane geometry question is to mark up the figure. Put as much of the information into the figure as you can. That's a good way to organize your thoughts. And that way you don't have to go back and forth between the figure and the question.

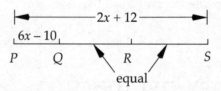

Now you can plan your attack. First subtract $\overline{PQ}$ from the whole length of $\overline{PS}$ to get $\overline{QS}$:

$$QS = PS - PQ = (2x + 12) - (6x - 10) = 2x + 12 - 6x + 10 = -4x + 22$$

Then, because R is the midpoint of $\overline{QS}$, you can divide QS by 2 to get QR and RS:

$$QR = RS = \frac{QS}{2} = \frac{-4x + 22}{2} = -2x + 11$$

Watch out! That matches choice (B), but it's the answer to the wrong question. What you're looking for is PR, so you have to add:

$$PR = PQ + QR = (6x - 10) + (-2x + 11) = 4x + 1$$

The answer is (E).

<blockquote>
MARK UP THE FIGURE

Try to put all the given information into the figure so that you can see everything at a glance and don't have to keep going back and forth between the question stem and the figure.
</blockquote>

Basic Traits of Triangles

Most Math 1 plane geometry questions are about closed figures: polygons and circles. And the test makers' favorite closed figure by far is the three-sided polygon; that is, the triangle. All three-sided polygons are interesting because they share so many characteristics, and certain special three-sided polygons—equilateral, isosceles, and right triangles—are interesting because of their special characteristics.

Let's look at the traits that all triangles share.

Sum of the interior angles: The three interior angles of any triangle add up to 180°.

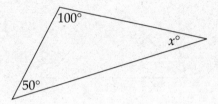

In the figure above, $x + 50 + 100 = 180$, so $x = 30$.

Measure of an exterior angle: The measure of an exterior angle of a triangle is equal to the sum of the measures of the remote interior angles.

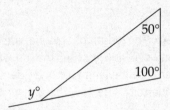

In the figure above, the measure of the exterior angle labeled $y°$ is equal to the sum of the measures of the remote interior angles: $y = 50 + 100 = 150$.

Sum of the exterior angles: The measures of the three exterior angles of any triangle add up to 360°.

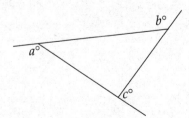

In the figure above, $a + b + c = 360$. (Note: In fact, the measures of the exterior angles of any polygon add up to 360°.)

<table>
<tr><td>

FIVE FACTS ABOUT TRIANGLES

1. Interior angles add up to 180°.

2. Exterior angle equals sum of remote interior angles.

3. Exterior angles add up to 360°.

4. Area of a triangle = $\frac{1}{2}$ (base)(height).

5. Each side is greater than the difference and less than the sum of the other two sides.

</td></tr>
</table>

Area formula: The general formula for the area of a triangle is always the same. The formula is this:

$$\text{Area of Triangle} = \frac{1}{2}(\text{base})(\text{height})$$

The height is the perpendicular distance between the side that's chosen as the base and the opposite vertex.

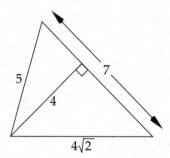

In the triangle above, 4 is the height when the 7 side is chosen as the base.

$$\text{Area of Triangle} = \frac{1}{2}(\text{base})(\text{height})$$

$$= \frac{1}{2}(7)(4) = 14$$

Triangle Inequality Theorem: The length of any one side of a triangle must be greater than the positive difference between, and less than the sum of, the lengths of the other two sides. For example, if it is given that the length of one side is 3 and the length of another side is 7, then the length of the third side must be greater than $7 - 3 = 4$ and less than $7 + 3 = 10$.

Similar Triangles

In Example 2 you're asked to express the area of one triangle in terms of the area of another.

Example 2

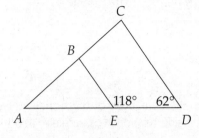

Figure 2

> **SIMILAR FIGURES**
>
> The area ratio between similar figures is the square of the side ratio.

In Figure 2, $AB = BC$. If the area of $\triangle ABE$ is x, what is the area of $\triangle ACD$?

(A) $x\sqrt{2}$

(B) $x\sqrt{3}$

(C) $2x$

(D) $3x$

(E) $4x$

You might wonder how you're supposed to find the area of $\triangle ACD$ when you're given no lengths you can use for a base or an altitude. The only numbers you have are the angle measures. They must be there for some reason—the test makers rarely provide superfluous information. In fact, because the two angle measures provided add up to 180°, they tell you that $\overline{BE}$ and $\overline{CD}$ are parallel. And that, in turn, tells you that $\triangle ABE$ is similar to $\triangle ACD$—because they have the same three angles.

Similar triangles are triangles that have the same shape: Corresponding angles are equal, and corresponding sides are proportional. In this case, because it's given that $AB = BC$, you know that AC is twice AB and that corresponding sides are in a ratio of 2:1. Each side of the larger triangle is twice the length of the corresponding side of the smaller triangle. That doesn't mean, however, that the ratio of the areas is also 2:1. In fact, the area ratio is the square of the side ratio, and the larger triangle has four times the area of the smaller triangle, so the answer is (E).

Alternative method: If you didn't see the similar triangles, or if you didn't know for sure how the area of the larger triangle is related to the area of the smaller triangle, you at least could have eliminated some answer choices based on appearances. Look at the figure and use your eyes to compare the areas. (We call this method eyeballing.) Doesn't it look as though the larger triangle has more than twice as much room inside it as the smaller triangle? That means that answer choices (A), (B), and (C) are all visibly too small. If you can narrow the choices down to two, it certainly pays to guess.

Eyeballing is never what you're supposed to do to answer a question, but if you don't see a better way, eyeballing's better than skipping.

> **EYEBALL THE FIGURE**
>
> You can assume that a figure is drawn to scale unless the problem says otherwise. So when you're stuck on a geometry question and don't know what else to do, see if you can at least use your eyes to eliminate a few answer choices as visibly too small or too big.

Three Special Triangle Types

Three special triangle types deserve extra attention:

- Isosceles triangles
- Equilateral triangles
- Right triangles

Be sure you know not just the definitions of these triangle types but, more importantly, their special characteristics: side relationships, angle relationships, and area formulas.

Isosceles triangle: An isosceles triangle is a triangle that has two equal sides. Not only are two sides equal, but the angles opposite the equal sides, called base angles, are also equal.

Equilateral triangle: An equilateral triangle is a triangle that has three equal sides. Since all the sides are equal, all the angles are also equal. All three angles in an equilateral triangle measure 60 degrees, regardless of the lengths of the sides. You can find the area of an equilateral triangle by dividing it into two 30-60-90 triangles, or you can use this formula in terms of the length of one side s:

$$\textbf{Area of Equilateral Triangle} = \frac{s^2\sqrt{3}}{4}$$

Right triangle: A right triangle is a triangle with a right angle. The two sides that form the right angle are called *legs*, and you can use them as the base and height to find the area of a right triangle.

$$\textbf{Area of Right Triangle} = \frac{1}{2}(\textbf{leg}_1)(\textbf{leg}_2)$$

Pythagorean theorem: If you know any two sides of a right triangle, you can find the third side by using the Pythagorean theorem:

$$(\textbf{leg}_1)^2 + (\textbf{leg}_2)^2 = (\textbf{hypotenuse})^2$$

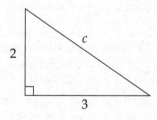

For example, if one leg is 2 and the other leg is 3, then

$$2^2 + 3^2 = c^2$$
$$c^2 = 4 + 9$$
$$c = \sqrt{13}$$

THREE SPECIAL TRIANGLE TYPES

1. **Isosceles:** Two equal sides and two equal angles.
2. **Equilateral:** Three equal sides and three 60° angles. If the length of one side is s, then

   Area of Equilateral Triangle $= \frac{s^2\sqrt{3}}{4}$.
3. **Right:** One right angle. You can use the legs to find the area:

   Area of Right Triangle $= \frac{1}{2}(\text{leg}_1)(\text{leg}_2)$.

PYTHAGOREAN THEOREM

For all right triangles, $(\text{leg}_1)^2 + (\text{leg}_2)^2 = (\text{hypotenuse})^2$.

Pythagorean triplet: A Pythagorean triplet is a set of integers that fit the Pythagorean theorem. The simplest Pythagorean triplet is (3, 4, 5). In fact, any integers in a 3:4:5 ratio make up a Pythagorean triplet. And there are many other Pythagorean triplets: (5, 12, 13); (7, 24, 25); (8, 15, 17); (9, 40, 41); all their multiples; and infinitely many more.

3-4-5 triangle: If a right triangle's leg-to-leg ratio is 3:4, or if the leg-to-hypotenuse ratio is 3:5 or 4:5, then it's a 3-4-5 triangle, and you don't need to use the Pythagorean theorem to find the third side. Just figure out what multiple of 3-4-5 it is.

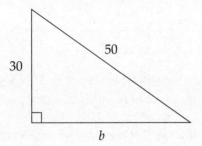

In the right triangle shown, one leg is 30 and the hypotenuse is 50. This is 10 times 3-4-5. The other leg is 40.

5-12-13 triangles: If a right triangle's leg-to-leg ratio is 5:12, or if the leg-to-hypotenuse ratio is 5:13 or 12:13, then it's a 5-12-13 triangle, and you don't need to use the Pythagorean theorem to find the third side. Just figure out what multiple of 5-12-13 it is.

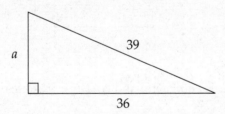

Here one leg is 36 and the hypotenuse is 39. This is 3 times 5-12-13. The other leg is 3 × 5 or 15.

45-45-90 triangles: The sides of a 45-45-90 triangle are in a ratio of $1:1:\sqrt{2}$.

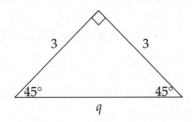

If one leg is 3, then the other leg is also 3, and the hypotenuse is equal to a leg times $\sqrt{2}$, or $3\sqrt{2}$.

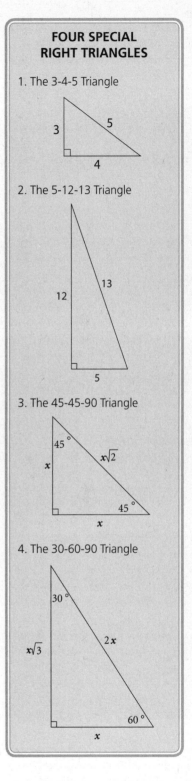

FOUR SPECIAL RIGHT TRIANGLES

1. The 3-4-5 Triangle

2. The 5-12-13 Triangle

3. The 45-45-90 Triangle

4. The 30-60-90 Triangle

30-60-90 triangles: The sides of a 30-60-90 triangle are in a ratio of $1 : \sqrt{3} : 2$. You don't need to use the Pythagorean theorem.

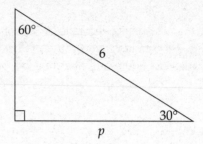

If the hypotenuse is 6, then the shorter leg is half that, or 3; and then the longer leg is equal to the short leg times $\sqrt{3}$, or $3\sqrt{3}$.

Example 3 includes one triangle that's equilateral and another that's both right and isosceles.

Example 3

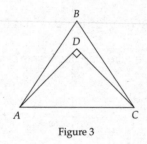

Figure 3

In Figure 3, $\triangle ABC$ is equilateral and $\triangle ADC$ is isosceles. If $AC = 1$, what is the distance from B to D?

(A)   0.286
(B)   0.318
(C)   0.333
(D)   0.366
(E)   0.383

To get the answer to this question, you need to know about equilateral, 45-45-90, and 30-60-90 triangles. If you drop an altitude from B through D, you will divide the equilateral triangle into two 30-60-90 triangles, and you will divide the right isosceles (or 45-45-90) triangle into two smaller right isosceles triangles:

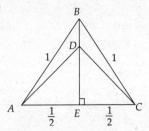

Using the side ratios for 30-60-90s and 45-45-90s, you know that $BE = \frac{\sqrt{3}}{2}$ and that $DE = \frac{1}{2}$.

Therefore, $BD = BE - DE = \frac{\sqrt{3}}{2} - \frac{1}{2} = \frac{\sqrt{3}-1}{2} \approx \frac{1.732-1}{2} = 0.366$.

The answer is (D).

Hidden Special Triangles

It happens a lot that the key to solving a geometry problem is to add a line segment or two to the figure. Often what results is one or more special triangles. The ability to spot, even to create, special triangles comes in handy in a question like Example 4.

Example 4

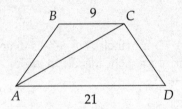

Figure 4

In Figure 4, the perimeter of isosceles trapezoid *ABCD* is 50. If *BC* = 9 and *AD* = 21, what is the length of diagonal *AC* ?

(A)   13

(B)   14

(C)   15

(D)   16

(E)   17

As you read the stem, you might wonder what an *isosceles trapezoid* is. If you'd never heard the term before, you still might have been able to extrapolate its meaning from what you know of isosceles triangles. Isosceles means "having two equal sides." When applied to a trapezoid, it tells you that the lengths of the two nonparallel sides—the legs—are equal. In this case that's AB and CD. If the total perimeter is 50, and the two marked sides add up to $21 + 9 = 30$, then the two unmarked sides split the difference of $50 - 30 = 20$. In other words, $AB = CD = 10$.

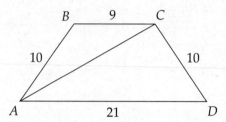

There aren't any special triangles yet. As so often happens, though, you can get some by constructing altitudes. Drop perpendiculars from points B and C, and you make two right triangles. The length 21 of side $\overline{AD}$ then gets split into 6, 9, and 6.

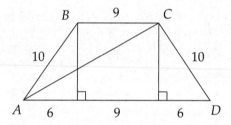

Now you can see that those right triangles are 3-4-5s (times 2) and that the height of the trapezoid is 8. Now look at the right triangle of which $\overline{AC}$ is the hypotenuse.

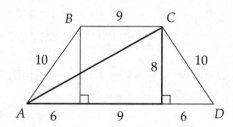

One leg is $6 + 9 = 15$, and the other leg is 8; therefore the hypotenuse AC is as follows.

$$\text{hypotenuse} = \sqrt{\left(\text{leg}_1\right)^2 + \left(\text{leg}_2\right)^2}$$
$$= \sqrt{15^2 + 8^2}$$
$$= \sqrt{225 + 64} = \sqrt{289} = 17$$

So $AC = 17$ and the answer is (E).

Special Quadrilaterals

The trapezoid is just one of five special quadrilaterals you need to be familiar with. As with triangles, there is some overlap among these categories, and some figures fit into none of these categories. Just as a 45-45-90 triangle is both right and isosceles, a quadrilateral with four equal sides and four right angles is not only a square but also a rhombus, a rectangle, and a parallelogram. It is wise to have a solid grasp of the definitions and special characteristics of these five quadrilateral types.

Trapezoids: A trapezoid is a four-sided figure with one pair of parallel sides and one pair of nonparallel sides.

$$\text{Area of Trapezoid} = \left(\frac{\text{base}_1 + \text{base}_2}{2} \right) \times \text{height}$$

Think of this formula as the average of the bases (the two parallel sides) times the height (the length of the perpendicular altitude).

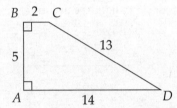

In the trapezoid *ABCD* above, you can use side $\overline{AB}$ for the height. The average of the bases is $\frac{2+14}{2} = 8$, so the area is 8 × 5, or 40.

Parallelograms: A parallelogram is a four-sided figure with two pairs of parallel sides. Opposite sides are equal. Opposite angles are equal. Consecutive angles add up to 180°.

Area of Parallelogram = base × height

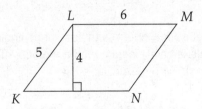

In parallelogram *KLMN* above, 4 is the height when $\overline{LM}$ or $\overline{KN}$ is used as the base. Base × height = 6 × 4 = 24.

Remember that to find the area of a parallelogram, you need the height, which is the perpendicular distance from the base to the opposite side. You can use a side of a parallelogram for the height only when the side is perpendicular to the base, in which case you have a rectangle.

Rectangles: A rectangle is a four-sided figure with four right angles. Opposite sides are equal. Diagonals are equal. The perimeter of a rectangle is equal to the sum of the lengths of the four sides, which is equal to 2(length + width).

Area of Rectangle = length × width

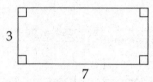

The area of a 7-by-3 rectangle is 7 × 3 = 21.

Rhombus: A rhombus is a four-sided figure with four equal sides.

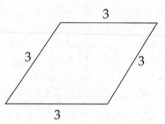

All four sides of the quadrilateral above have the same length, so it's a rhombus. A rhombus is also a parallelogram, so to find the area of a rhombus, you need its height. The more a rhombus "leans over," the smaller the height and therefore the smaller the area. The shape with the maximum area for a rhombus of a certain perimeter is a rhombus that has each pair of adjacent sides perpendicular, in which case you have a square.

Square: A square is a four-sided figure with four right angles and four equal sides. A square is also a rectangle, a parallelogram, and a rhombus. The perimeter of a square is equal to 4 times the length of one side.

Area of Square = (side)^2

The square above, with sides of length 5, has an area of $5^2 = 25$.

Polygons—Perimeter and Area

The test makers like to write problems that combine the concepts of perimeter and area. What you need to remember is that perimeter and area are not directly related. In Example 5, for instance, you have two figures with the same perimeter, but that doesn't mean they have the same area.

Example 5

A square and a regular hexagon have the same perimeter. If the area of the square is 2.25, what is the area of the hexagon?

(A) 2.250
(B) 2.598
(C) 2.838
(D) 3.464
(E) 3.375

The way to get started with this question is to sketch what's described in the question. A square of area 2.25 has sides each of length $\sqrt{2.25} = 1.5$. So the perimeter of the square is $4(1.5) = 6$. Since that's also the perimeter of the regular hexagon, and a regular hexagon has six equal sides, the length of each side of the hexagon is 1.

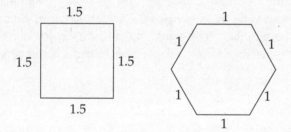

Now the problem is one of finding the area of a regular hexagon of side length 1. The fastest way to do that would be to use the formula, if you know it. If the length of one side is s:

$$\text{Area of Hexagon} = \frac{3s^2\sqrt{3}}{2}$$

This formula is not one the test makers expect you to know—there's always a way around it—but if you like formulas and you're good at memorizing them, it can only help. Let's proceed, however, as if we didn't know the formula. Another way to go about finding this area is to add a line segment or two to the figure and divide it up into more familiar shapes. You could, for example, draw in three diagonals and turn the hexagon into six equilateral triangles of side 1.

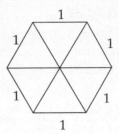

Each of those six triangles has base 1 and height $\dfrac{\sqrt{3}}{2}$, and therefore,

$$\textbf{Area of one triangle} = \frac{1}{2}(\text{base})(\text{height}) = \frac{1}{2}(1)\left(\frac{\sqrt{3}}{2}\right) = \frac{\sqrt{3}}{4}$$

The area of the hexagon is 6 times that.

$$\textbf{Area of hexagon} = 6\left(\frac{\sqrt{3}}{4}\right) = \frac{3\sqrt{3}}{2} \approx 2.598$$

The answer is (B).

Circles—Four Formulas

After the triangle, the test makers' favorite plane geometry figure is the circle. Circles don't come in as many varieties as triangles do. In fact, all circles are similar—they're all the same shape. The only difference among them is size. So you don't have to learn to recognize types or remember names. All you have to know about circles is how to find four things:

- Circumference
- Length of an arc
- Area
- Area of a sector

You could think of the task as one of memorizing four formulas, but you'll be better off in the end if you have some idea of where the arc and sector formulas come from and how they are related to the circumference and area formulas.

Circumference: Circumference is a measurement of length. You could think of it as the perimeter: It's the total distance around the circle. If the radius of the circle is *r*,

$$\textbf{Circumference} = 2\pi r$$

Since the diameter is twice the radius, you can easily express the formula in terms of the diameter *d*:

$$\textbf{Circumference} = \pi d$$

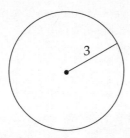

In the circle above, the radius is 3, so the circumference is $2\pi(3) = 6\pi$.

Length of an arc: An arc is a piece of the circumference. If n is the degree measure of the arc's central angle, then the formula is:

$$\text{Length of an Arc} = \left(\frac{n}{360}\right)(2\pi r)$$

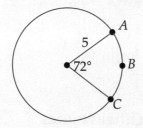

In the figure above, the radius is 5, and the measure of the central angle is 72°. The arc length is $\frac{72}{360}$ or $\frac{1}{5}$ of the circumference:

$$\left(\frac{72}{360}\right)(2\pi)(5) = \left(\frac{1}{5}\right)(10\pi) = 2\pi$$

Area: The area of a circle is usually found using this formula in terms of the radius r:

$$\text{Area of a Circle} = \pi r^2$$

The area of the circle above is $\pi(4)^2 = 16\pi$.

Area of a sector: A sector is a piece of the area of a circle. If n is the degree measure of the sector's central angle, then the area formula is

$$\text{Area of a Sector} = \left(\frac{n}{360}\right)\left(\pi r^2\right)$$

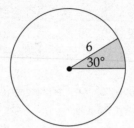

In the figure above, the radius is 6, and the measure of the sector's central angle is 30°. The sector has $\frac{30}{360}$ or $\frac{1}{12}$ of the area of the circle:

$$\left(\frac{30}{360}\right)(\pi)(6^2) = \left(\frac{1}{12}\right)(36\pi) = 3\pi$$

Circles Combined with Other Figures

Some of the most challenging plane geometry questions are those that combine circles with other figures. Consider Example 6:

Example 6

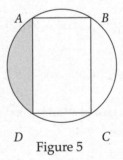

Figure 5

In Figure 5, rectangle $ABCD$ is inscribed in a circle. If the radius of the circle is 1 and $AB = 1$, what is the area of the shaded region?

(A)   0.091
(B)   0.285
(C)   0.614
(D)   0.705
(E)   0.732

Once again, the key here is to add to the figure. And in this case, as so often when there's a circle, what you should add is radii. The equilateral triangles tell you that the central angles are 60° and 120°.

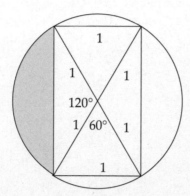

The shaded region is what's left of the 120° sector after you subtract the triangle with the 120° vertex angle.

To find the area of the shaded region, find the areas of the sector and triangle, then subtract. The sector is exactly one-third of the circle (because 120° is one-third of 360°), so

$$\text{Area of sector} = \frac{1}{3}\pi r^2 = \frac{1}{3}\pi(1)^2 = \frac{\pi}{3} \approx 1.047$$

You can divide the triangle into two 30-60-90s:

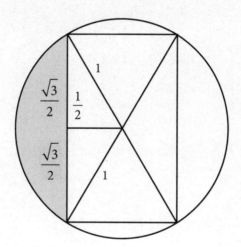

The area of each 30-60-90 triangle is $\frac{1}{2}\left(\frac{1}{2}\right)\left(\frac{\sqrt{3}}{2}\right) = \frac{\sqrt{3}}{8}$, so the area of the triangle with the 120° vertex is twice that, or $\frac{\sqrt{3}}{4} \approx 0.433$.

The shaded area, then, is about 1.047 − 0.433 = 0.614. The answer is (C).

Now that you've reviewed all the relevant plane geometry facts and formulas, seen some of the test makers' favorite plane geometry situations, and learned a few good Kaplan strategies, it's time to put yourself to the test again. If you were disappointed in your performance on the Plane Geometry Diagnostic Test, here's your chance to make up for it.

THINGS TO REMEMBER:

Five Facts about Triangles

1. Interior angles add up to 180°.
2. Exterior angle equals sum of remote interior angles.
3. Exterior angles add up to 360°.
4. Area of a triangle = $\frac{1}{2}$(base)(height).
5. Each side is greater than the difference and less than the sum of the other two sides.

Similar Figures

The area ratio between similar figures is the square of the side ratio.

Three Special Triangle Types

1. Isosceles: Two equal sides and two equal angles.
2. Equilateral: Three equal sides and three 60° angles. If the length of one side is s, then

   Area of Equilateral Triangle =

   $$\frac{s^2\sqrt{3}}{4}$$

3. Right: One right angle. You can use the legs to find the area:

   Area of Right Triangle =

   $$\frac{1}{2}(\text{leg}_1)(\text{leg}_2)$$

Pythagorean Theorem

For all right triangles,

$(\text{leg}_1)^2 + (\text{leg}_2)^2 = (\text{hypotenuse})^2$

Four Special Right Triangles

1. The 3-4-5 Triangle

2. The 5-12-13 Triangle

3. The 45-45-90 Triangle

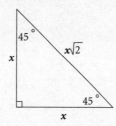

4. The 30-60-90 Triangle

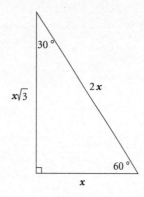

Five Special Quadrilateral Types

1. Trapezoid:

   Area of Trapezoid = $\dfrac{base_1 + base_2}{2} \times$ height

2. Parallelogram:

   Area of Parallelogram = base × height

3. Rectangle:

   Perimeter of Rectangle = 2(length + width)

   Area of Rectangle = length × width

4. Rhombus (four equal sides):

   Perimeter of Rhombus = 4 × side

   Area of Rhombus = base × height

5. Square (four right angles and four equal sides):

   Perimeter of Square = 4 × side

   Area of Square = (side)^2

Polygon Angles

For any polygon of n sides:

   Sum of Interior Angles = $(n - 2) \times 180°$

   Sum of Exterior Angles = $360°$

Four Circle Formulas

Circumference = $2\pi r = \pi d$

Length of an Arc = $\left(\dfrac{n}{360}\right)(2\pi r)$

Area of a Circle = πr^2

Area of a Sector = $\left(\dfrac{n}{360}\right)(\pi r^2)$

PLANE GEOMETRY FOLLOW-UP TEST

6 Questions (8 Minutes)

Directions: Solve the following problems. Fill in the oval corresponding to the best answer choice in the grid to the right of each question. (Answers and explanations begin on page 88.)

DO YOUR FIGURING HERE

1. In Figure 1, the ratio of *AB* to *BC* is 7 to 5. If *AC* = 1, what is the distance from *A* to the midpoint of $\overline{BC}$?

    (A) $\dfrac{5}{8}$

    (B) $\dfrac{2}{3}$

    (C) $\dfrac{17}{24}$

    (D) $\dfrac{3}{4}$

    (E) $\dfrac{19}{24}$ Ⓐ Ⓑ Ⓒ Ⓓ Ⓔ

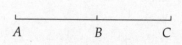

Figure 1

2. In △*PRS* in Figure 2, $\overline{RT}$ is the altitude to side $\overline{PS}$, and $\overline{QS}$ is the altitude to side $\overline{PR}$. If *RT* = 7, *PR* = 8, and *QS* = 9, what is the length of side $\overline{PS}$?

    (A)   5.14
    (B)   6.22
    (C)   7.87
    (D)   10.29
    (E)   13.44   Ⓐ Ⓑ Ⓒ Ⓓ Ⓔ

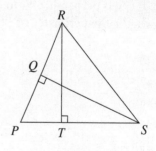

Figure 2

3. In Figure 3, $\overline{QS}$ and $\overline{PT}$ are parallel, and the lengths of segments $\overline{PQ}$ and $\overline{QR}$ are as marked. If the area of $\triangle QRS$ is x, what is the area of $\triangle PRT$ in terms of x ?

    (A) $\dfrac{3x}{2}$

    (B) $\dfrac{9x}{4}$

    (C) $\dfrac{5x}{2}$

    (D) $3x$

    (E) $\dfrac{25x}{4}$ Ⓐ Ⓑ Ⓒ Ⓓ Ⓔ

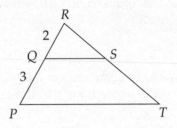

Figure 3

4. In Figure 4, if $AB = 2$, what is the area of $\triangle ABC$?

    (A) 2.45
    (B) 2.73
    (C) 3.86
    (D) 4.89
    (E) 5.46   Ⓐ Ⓑ Ⓒ Ⓓ Ⓔ

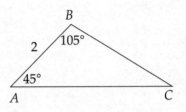

Figure 4

5. In Figure 5, the area of parallelogram $PQRS$ is 24 and $QR = 6$. If $\overline{QT}$ is perpendicular to $\overline{PS}$ and if T is the midpoint of $\overline{PS}$, what is the perimeter of $PQRS$?

    (A) 20
    (B) 22
    (C) 24
    (D) 26
    (E) 28   Ⓐ Ⓑ Ⓒ Ⓓ Ⓔ

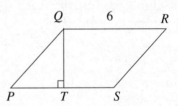

Figure 5

6. In Figure 6, points A, B, and C lie on the circumference of the circle centered at O. If $\angle OAB$ measures 50° and $\angle BCO$ measures 60°, what is the measure of $\angle AOC$?

    (A) 110°
    (B) 120°
    (C) 130°
    (D) 140°
    (E) 150°   Ⓐ Ⓑ Ⓒ Ⓓ Ⓔ

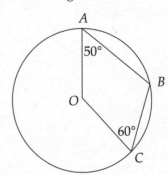

Figure 6

**Turn the page
for answers and explanations
to the Follow-Up Test.**

FOLLOW-UP TEST—ANSWERS AND EXPLANATIONS

1. E

Mark up the figure:

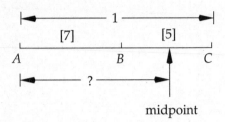

The brackets around the 7 and 5 are meant to show that they're proportions, not actual lengths. Since the ratio of AB to BC is 7 to 5, you can say that 7 parts out of 12 are in AB and 5 parts out of 12 are in BC.

Because the whole length of $\overline{AC}$ is 1, $AB = \dfrac{7}{12}$ and $BC = \dfrac{5}{12}$. The midpoint of $\overline{BC}$ divides it in half, so each half length is half of $\dfrac{5}{12}$, which is $\dfrac{5}{24}$. The length you're looking for is $\dfrac{7}{12} + \dfrac{5}{24} = \dfrac{14}{24} + \dfrac{5}{24} = \dfrac{19}{24}$.

2. D

Put the given lengths into the figure:

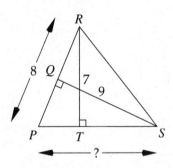

The key is realizing that whatever base-altitude pair you take for the same triangle, you'll get the same area. One-half the product of PS and RT is the same as one-half the product of PR and QS. Or, ignoring the one-halfs, you can just say that the products are equal.

$$\frac{1}{2}(PS)(RT) = \frac{1}{2}(PR)(QS)$$

$$(PS)(RT) = (PR)(QS)$$

$$(PS)(7) = (8)(9)$$

$$PS = \frac{(8)(9)}{7} = \frac{72}{7} = 10.29$$

3. E

The only information that's in the question and not in the figure is that $\overline{QS}$ and $\overline{PT}$ are parallel and that the area of $\triangle QRS$ is x. That $\overline{QS}$ and $\overline{PT}$ are parallel tells you that triangles PRT and QRS are similar—they have the same angles. Because the triangles are similar, the sides are proportional. Because the ratio of PR to QR is $\dfrac{5}{2}$, the ratio of any pair of corresponding sides will also be $\dfrac{5}{2}$. But that's not the ratio of the areas. Remember that the area ratio between similar figures is the square of the side ratio. Here the side ratio is $\dfrac{5}{2}$, so the area ratio is $\left(\dfrac{5}{2}\right)^2 = \dfrac{25}{4}$. If the area of the small triangle is x, then the area of the large one is $\dfrac{25x}{4}$.

4. B

Drop an altitude, and you'll reveal two hidden special triangles:

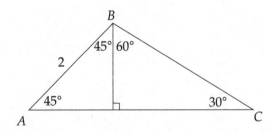

Now you can use the known side ratios of 45-45-90 and 30-60-90 triangles to get all the lengths you need. If the hypotenuse of a 45-45-90 is 2, then each leg is $\dfrac{2}{\sqrt{2}} = \sqrt{2}$:

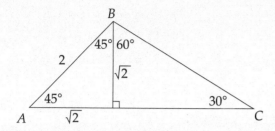

Now that you know the short leg of the 30-60-90, you can multiply that by $\sqrt{3}$ to find that the longer leg is $\sqrt{6}$.

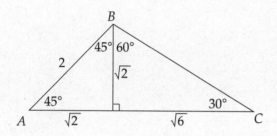

Now you have the base and the height. The base is $\sqrt{2}+\sqrt{6}$, and the height is $\sqrt{2}$, so

Area of a triangle $= \dfrac{1}{2}$ *(base)(height)*

$$= \frac{1}{2}\left(\sqrt{2}+\sqrt{6}\right)\left(\sqrt{2}\right)=\frac{1}{2}\left(2+\sqrt{12}\right)$$

$$= \frac{1}{2}\left(2+2\sqrt{3}\right)=1+\sqrt{3}\approx 2.73$$

5.   B

Because *PQRS* is a parallelogram, opposite sides are equal and *PS* = 6. Midpoint *T* divides that 6 into two 3s. The area of the parallelogram, which is equal to the base times the height, is given as 24, and because the base is 6, altitude $\overline{QT}$ must be 4. Put all this into the figure, and the hidden special triangle reveals itself.

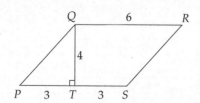

Now you can see that Δ*PQT* is a 3-4-5 right triangle and that *PQ* = 5. Opposite sides are equal, so *RS* = 5, too, and the perimeter is 5 + 6 + 5 + 6 = 22.

6.   D

As is so often true when circles are combined with other figures, the key to solving this question is to draw in a radius:

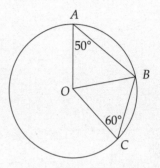

All the radii of a circle are equal, so within each of the two triangles you just created, the angles opposite the radii are equal:

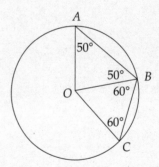

Now that you know two angles in each triangle, you can figure out the third:

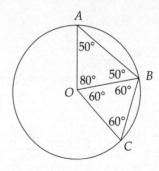

The angle you're looking for measures 80° + 60° = 140°.

Chapter 6: **Solid Geometry**

- Volume and surface area
- Test makers' favorites
- Picturing solids

Math 1 includes a few solid geometry questions. But this is not a big and vital category.

HOW TO USE THIS CHAPTER

You may already know all the solid geometry you need. Find out by taking the Solid Geometry Diagnostic Test. The five questions on the Diagnostic Test are typical of those on the Mathematics subject tests. Check your answers using the answer key following the test. No matter how you score, don't worry! The answer key also shows where to find a detailed explanation for each question. The "Find Your Study Plan" section that follows the test will suggest next steps based on your performance on the Diagnostic.

Find Your Level

How you use this chapter really depends on which test you're taking and how much time you have to prep. Find your level and pace below.

Standard Plan. Take the Solid Geometry Diagnostic Test to find out how much you already know about solid geometry. Then read the rest of the chapter and do the Follow-Up Test at the end.

> *Shortcut: Take the Solid Geometry Diagnostic Test and check your answers. If you can answer most of the questions correctly, you should skip this chapter.*

Panic Plan. Skip this chapter.

SOLID GEOMETRY DIAGNOSTIC TEST

5 Questions (6 Minutes)

Directions: Solve the following problems. Fill in the oval corresponding to the best answer choice in the grid to the right of each question. (Answers are on page 95.)

> **Reference Information:** Use the following formulas as needed.
>
> **Right circular cone:** If r = radius and h = height, then Volume = $\frac{1}{3}\pi r^2 h$; and if c = circumference of the base and ℓ = slant height, then Lateral Area = $\frac{1}{2}c\ell$.
>
> **Sphere:** If r = radius, then Volume = $\frac{4}{3}\pi r^3$ and Surface Area = $4\pi r^2$.
>
> **Pyramid:** If B = area of the base and h = height, then Volume = $\frac{1}{3}Bh$.

1. A right circular cone and a sphere have equal volumes. If the radius of the base of the cone is $2x$ and the radius of the sphere is $3x$, what is the height of the cone in terms of x ?

   (A) x (B) $\frac{3x}{2}$ (C) $\frac{4x}{3}$ (D) $20x$ (E) $27x$

   Ⓐ Ⓑ Ⓒ Ⓓ Ⓔ

2. In the rectangular solid in Figure 1, what is the distance from vertex A to vertex B ?

   (A) $\sqrt{65}$
   (B) $7\sqrt{2}$
   (C) 9
   (D) 11
   (E) 15

   Ⓐ Ⓑ Ⓒ Ⓓ Ⓔ

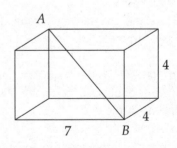

Figure 1

KAPLAN

...ght uniform solid are ...and 5. If the volume of ...face area?

Ⓐ Ⓑ Ⓒ Ⓓ Ⓔ

...ted 360° about side $\overline{BC}$, ...solid?

(D)   37.70

(E)   56.55

Ⓐ Ⓑ Ⓒ Ⓓ Ⓔ

5. The maximum possible number of identical rectangular blocks are placed inside a rectangular carton. The rectangular blocks each have a length of 7, a width of 5, and a height of 4, and the rectangular carton has inside dimensions that are a length of 16, a width of 15, and a height of 14. What is the total surface area of the rectangular blocks that are inside the rectangular carton?

(A)   1,348

(B)   1,992

(C)   2,656

(D)   3,360

(E)   3,984

Ⓐ Ⓑ Ⓒ Ⓓ Ⓔ

DO YOUR FIGURING HERE

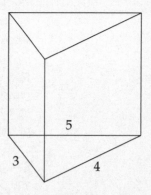

Figure 2

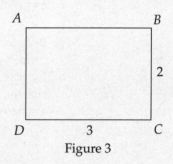

Figure 3

Find Your Study Plan

The answer key on the following page shows where in this chapter to find explanations for the questions you missed. Here's how you should proceed based on your Diagnostic Test score.

5: Superb! You have a solid grasp of the essentials of solid geometry, and if you're pressed for time, you might consider skipping this chapter. If you want to try your hand at a few more solid geometry questions, go straight to the Follow-Up Test at the end of this chapter.

3–4: Good! If you can get three or four of these relatively difficult questions right, then you have a decent understanding of solid geometry. If you're pressed for time, you might consider skipping this chapter. You might want to look over the parts of this chapter that deal with the one or two questions you didn't get right. If you just want to try your hand at a few more solid geometry questions, go straight to the Follow-Up Test at the end of this chapter.

0–2: Solid geometry is not your forte right now, but since it accounts for only three or four questions on the test, it may not be worth worrying about. If you're pressed for time and are confident with the other test topics, you might consider skipping this chapter. But if you have the time, you should read the rest of this chapter, study the examples, and then try the Follow-Up Test at the end of the chapter.

SOLID GEOMETRY TEST TOPICS

The questions in the Solid Geometry Diagnostic Test are typical of those on the Mathematics subject test. In this chapter we'll use these questions to review the solid geometry you're expected to know. We will also use these questions to demonstrate some effective problem-solving techniques, alternative methods, and test-taking strategies that apply to solid geometry questions.

Five Formulas You Don't Need to Memorize

Some people find solid geometry intimidating because of all those formulas. On the Mathematics subject test, however, some of the scariest formulas are given to you in the directions. It's good to know what formulas are included in the directions so you don't waste time memorizing them—and so that you won't forget to memorize all the relevant formulas that are not included in the directions. These are the formulas that are printed in the directions.

Lateral area of a cone: Given base circumference c and slant height ℓ,

$$\textbf{Lateral Area of Cone} = \frac{1}{2}c\ell$$

The lateral area of a cone is the area of the part that extends from the vertex to the circular base. It does not include the circular base.

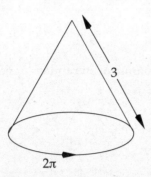

For example, in the figure above, $c = 2\pi$ and $\ell = 3$, so

$$\text{Lateral Area} = \frac{1}{2}(2\pi)(3) = 3\pi$$

Volume of a cone: Given base radius r and height h,

$$\textbf{Volume of Cone} = \frac{1}{3}\pi r^2 h$$

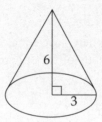

For example, in the figure above, $r = 3$ and $h = 6$, so

$$\text{Volume} = \frac{1}{3}\pi(3^2)(6) = 18\pi$$

Surface area of a sphere: Given radius r,

$$\textbf{Surface Area of Sphere} = 4\pi r^2$$

For example, if the radius of a sphere is 2, then

$$\textbf{Surface Area} = 4\pi(2^2) = 16\pi$$

Volume of a sphere: Given radius r,

$$\textbf{Volume of Sphere} = \frac{4}{3}\pi r^3$$

For example, if the radius of a sphere is 2, then:

$$\text{Volume} = \frac{4}{3}\pi(2)^3 = \frac{32\pi}{3}$$

Volume of a pyramid: Given base area B and height h,

$$\textbf{Volume of Pyramid} = \frac{1}{3}\textbf{\textit{Bh}}$$

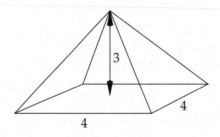

For example, in the figure above, $h = 3$ and, since the base is a square, $B = 16$, then:

$$\text{Volume} = \frac{1}{3}(16)(3) = 16$$

Here's a question that uses some of these formulas.

Example 1

A right circular cone and a sphere have equal volumes. If the radius of the base of the cone is $2x$ and the radius of the sphere is $3x$, what is the height of the cone in terms of x ?

(A) x (B) $\dfrac{3x}{2}$ (C) $\dfrac{4x}{3}$ (D) $20x$ (E) $27x$

As you can see, this question is hard enough even with the formulas provided. This is no mere matter of plugging values into a formula and cranking out the answer. This question is more algebraic than that and takes a little thought. It's really a word problem. It describes in words a mathematical situation (in this case, geometric) that can be translated into algebra. The pivot in this situation is that the cone and sphere have equal volumes. You're looking for the height h in terms of x, and fortunately you can express both volumes in terms of those two variables. Be careful. Both formulas include r, but they're not the same r's. In the case of the cone, $r = 2x$, but in the case of the sphere, $r = 3x$:

$$\text{Volume of cone} = \frac{1}{3}\pi r^2 h = \frac{1}{3}\pi(2x)^2 h = \frac{4}{3}\pi x^2 h$$

$$\text{Volume of sphere} = \frac{4}{3}\pi r^3 = \frac{4}{3}\pi(3x)^3 = 36\pi x^3$$

Now write an equation that says that the expressions for the two volumes are equal to each other, and solve for h:

$$\frac{4}{3}\pi x^2 h = 36\pi x^3$$

$$\pi x^2 h = \frac{3}{4}\left(36\pi x^3\right)$$

$$\pi x^2 h = 27\pi x^3$$

$$h = \frac{27\pi x^3}{\pi x^2} = 27x$$

And so the answer is (E).

The direct way to do this question is the algebraic way. But as you can see, the algebra is quite convoluted, and the matter of the different r's can be confusing. There's another, less sophisticated way to do this question—*pick numbers*. All the given measures are in terms of x, and all the answer choices are in terms of x, so to make things simpler, you could just pick a number for x, plug it into the question, and see what you get. Pick a number that's easy to work with. Here you could even pick $x = 1$ because it's clear that when $x = 1$, all the answer choices have different values.

If $x = 1$, then the radius of the base of the cone is 2, and the radius of the sphere is 3.

$$\text{Volume of cone} = \frac{1}{3}\pi r^2 h = \frac{1}{3}\pi (2)^2 h = \frac{4}{3}\pi h$$

$$\text{Volume of sphere} = \frac{4}{3}\pi r^3 = \frac{4}{3}\pi (3)^3 = 36\pi$$

So what does h have to be to give the cone a volume of 36π?

$$\frac{4}{3}\pi h = 36\pi$$

$$\frac{4}{3}h = 36$$

$$h = 36 \times \frac{3}{4} = 27$$

When $x = 1$, h turns out to be 27. Now plug $x = 1$ into the answer choices, and you'll find that only (E) gives you 27. This alternative method is still somewhat algebraic, but this way the algebra's a lot less convoluted.

> ### PICK NUMBERS
>
> When the answer choices are algebraic expressions, it is often quickest to pick numbers for the unknowns, plug those numbers into the stem and see what you get, and then plug those same numbers into the answer choices to find matches.
>
> *Warning:* When you pick numbers, you have to check all the answer choices. Sometimes more than one works with the number(s) you pick, in which case you have to pick numbers again.

> ### BRUSH UP ON YOUR ALGEBRA
>
> You can't escape algebra! The Mathematics subject tests are both full of algebra. Make sure your algebra skills are in peak condition by test day.

KAPLAN

The Test Makers' Favorite Solid

The test makers' favorite solid is the rectangular solid. That's the official geometric term for a box that has 6 rectangular faces and 12 edges that meet at right angles at 8 vertices.

SURFACE AREA OF A RECTANGULAR SOLID

If the length is ℓ, the width is w, and the height is h, the formula is

Surface Area = $2\ell w + 2wh + 2\ell h$

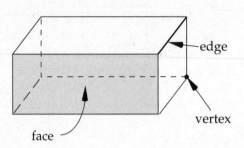

The surface area of a rectangular solid is simply the sum of the areas of the faces. That's what the formula "Surface Area = $2\ell w + 2wh + 2\ell h$" says. If the length is ℓ, the width is w, and the height is h, then two rectangular faces have area ℓw, two have area wh, and two have area ℓh. The total surface area is the sum of those three pairs of areas.

Instead of the surface area, you may be asked to find the distance between opposite vertices of a rectangular solid. Look at Example 2.

Example 2

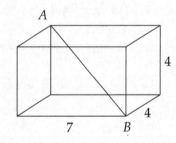

Figure 1

In the rectangular solid in Figure 1, what is the distance from vertex A to vertex B ?

(A) $\sqrt{65}$ (B) $7\sqrt{2}$ (C) 9   (D) 11   (E) 15

One way to find this distance is to apply the Pythagorean theorem twice. First plug the dimensions of the base into the Pythagorean theorem to find the diagonal of the base:

$$\text{Diagonal of base} = \sqrt{4^2 + 7^2} = \sqrt{16 + 49} = \sqrt{65}$$

Notice that the base diagonal combines with an edge and with the segment whose length you're looking for to form a right triangle:

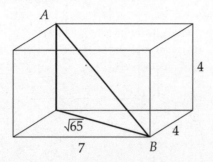

So you can plug the base diagonal and the height into the Pythagorean theorem to find *AB*:

$$AB = \sqrt{\left(\sqrt{65}\right)^2 + 4^2} = \sqrt{65 + 16} = \sqrt{81} = 9$$

The answer is (C).

Another way to find this distance is to use the formula below, which you could say is just the Pythagorean theorem taken to another dimension. If the length is ℓ, the width is w, and the height is h, the formula for the diagonal is

$$\textbf{Distance} = \sqrt{\boldsymbol{\ell^2 + w^2 + h^2}}$$

> **DISTANCE BETWEEN OPPOSITE VERTICES OF A RECTANGULAR SOLID**
>
> Distance = $\sqrt{\ell^2 + w^2 + h^2}$

Uniform Solids

A rectangular solid is one type of *uniform solid*. A uniform solid is what you get when you take a plane and move it, without tilting it, through space. Here are some uniform solids.

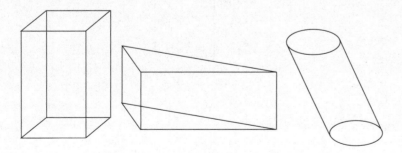

The way these solids are drawn, the top and bottom faces are parallel and congruent. These faces are called the *bases*. You can think of each of these solids as the result of sliding the base through space. The perpendicular distance through which the base slides is called the *height*.

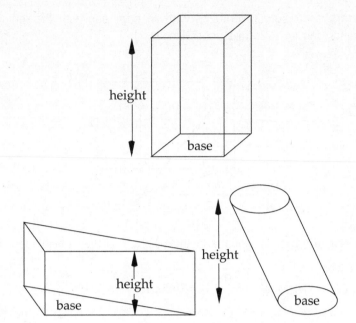

In every one of the above cases—indeed, in the case of *any* uniform solid—the volume is equal to the area of the base times the height. So, you can say that for any uniform solid, given the area of the base *B* and the height *h*,

Volume of a Uniform Solid = *Bh*

Volume of a rectangular solid: A rectangular solid is a uniform solid whose base is a rectangle and whose height is perpendicular to its base. Given the length ℓ, width *w*, and height *h*, the area of the base is ℓw, and so the volume formula is this:

Volume of a Rectangular Solid = ℓwh

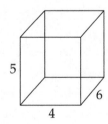

The volume of a 4-by-5-by-6 box is

$$4 \times 5 \times 6 = 120$$

Volume of a cube: A cube is a rectangular solid with length, width, and height all equal. If *s* is the length of an edge of a cube, the volume formula is

Volume of a Cube = s^3

**VOLUME
FORMULAS FOR
UNIFORM SOLIDS**

Volume of a Uniform
Solid = *Bh*

Volume of a Rectangular
Solid = ℓwh

Volume of a Cube = s^3

Volume of a Cylinder = $\pi r^2 h$

The volume of this cube is $2^3 = 8$.

Volume of a cylinder: A cylinder is a uniform solid whose base is a circle. Given base radius r and height h, the area of the base is πr^2, and so the volume formula is this:

$$\textbf{Volume of a Cylinder} = \boldsymbol{\pi r^2 h}$$

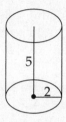

In the cylinder above, $r = 2$ and $h = 5$, so

$$\text{Volume} = \pi(2^2)(5) = 20\pi$$

Example 3 gives you the volume of a uniform solid and asks for the surface area.

Example 3

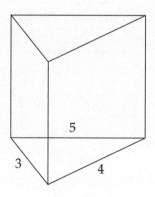

Figure 2

In Figure 2, the bases of the right uniform solid are triangles with sides of lengths 3, 4, and 5. If the volume of the solid is 30, what is the total surface area?

(A) 17   (B) 30   (C) 36   (D) 60   (E) 72

The surface area is the sum of the areas of the faces. To find the areas of the faces, you need to figure out what kinds of polygons they are so that you'll know what formulas to use. Start with the bases, which are said to be "triangles with sides of lengths 3, 4, and 5." If these side lengths don't ring a bell in your head, then you'd better go back to chapter 5 and bone up on your special triangles. This is a 3-4-5 triangle, which means that it's a right triangle, which means that you can use the legs as the base and height to find the area:

$$\text{Area of Right Triangle} = \frac{1}{2}\left(\text{leg}_1\right)\left(\text{leg}_2\right) = \frac{1}{2}(3)(4) = 6$$

That's the area of each of the bases. The other three faces are rectangles. To find their areas, you need first to determine the height of the solid. If the area of the base is 6, and the volume is 30, then

$$\text{Volume} = Bh$$
$$30 = 6h$$
$$5 = h$$

So the areas of the three rectangular faces are 3 × 5 = 15, 4 × 5 = 20, and 5 × 5 = 25. The total surface area, then, is 6 + 6 + 15 + 20 + 25 = 72. The answer is (E).

Picturing Solids

You might have thought that solid geometry questions are difficult because of the formulas. As you have seen, however, using formulas is by definition routine. It's the nonroutine problems that can be the most challenging. Those are the solid geometry problems that require you to visualize, like Example 4.

Example 4

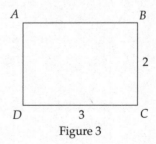

Figure 3

If the rectangle in Figure 3 is rotated 360° about side $\overline{BC}$, what is the volume of the resulting solid?

(A) 12.00   (B) 18.00   (C) 28.27   (D) 37.70   (E) 56.55

Can you visualize—picture in your mind—what the resulting solid looks like? It can be a little difficult to sketch 3-D geometry. You may be better off if you can just "see" it in your head. It'll look something like this.

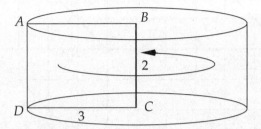

It's a cylinder with base radius 3 and height 2. So now all you have to do is plug $r = 3$ and $h = 2$ into the formula:

$$\text{Volume of Cylinder} = \pi r^2 h = \pi (3^2)(2) = 18\pi \approx 56.55$$

The answer is (E). Applying the formula was the easy part. Visualizing the cylinder and figuring out what's r and what's h was the challenging part. A typical Mathematics subject test will include one or two questions that entail visualizing. Here's another one in Example 5.

Example 5

The maximum possible number of identical rectangular blocks are placed inside a rectangular carton. The rectangular blocks each have a length of 7, a width of 5, and a height of 4, and the rectangular carton has inside dimensions that are a length of 16, a width of 15, and a height of 14. What is the total surface area of the rectangular blocks that are inside the rectangular carton?

  (A) 1,348   (B) 1,992   (C) 2,656   (D) 3,360   (E) 3,984

Notice that each of the dimensions 7, 5, and 4 of each rectangular block is a factor of one of the inside dimensions 16, 15, and 14 of the rectangular carton. Specifically, 7 is a factor of 14 (14 = 2 × 7), 5 is a factor of 15 (15 = 3 × 5), and 4 is a factor of 16 (16 = 4 × 4). So you can place the blocks in the carton by placing the dimension 7 of the blocks along the dimension 14 of the carton, the dimension 5 of the blocks along the dimension 15 of the carton, and the dimension 4 of the blocks along the dimension 16 of the carton. By placing the blocks in the carton in this way, you will completely fill up the rectangular carton. The number of times you can place the dimension 7 of a block along the dimension 14 of the carton is $\frac{14}{7}$, or 2. The number of times you can place the dimension 5 of a block along the dimension 15 of the carton is $\frac{15}{5}$, or 3. The number of times you can place the dimension 4 of a block along the dimension 16 of the carton is $\frac{16}{4}$, or 4. So it is possible to fill the volume of the carton completely with 2 × 3 × 4 = 6 × 4 = 24 rectangular blocks. Thus, the maximum number of rectangular blocks that can be placed in the carton is 24.

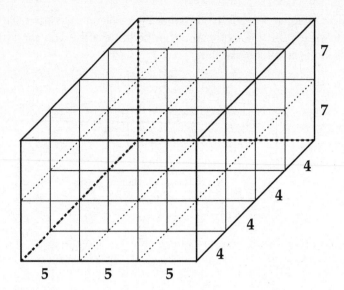

Now all that remains is to find the surface area of one rectangular block and multiply that by 24. The formula for the surface area S of a rectangular solid having a length ℓ, a width w, and a height h is

$S = 2\ell w + 2\ell h + 2wh$

Here, for each rectangular block, $\ell = 7$, $w = 5$, and $h = 4$. So the surface area of one rectangular block is

$2 \times 7 \times 5 + 2 \times 7 \times 4 + 2 \times 5 \times 4 = 70 + 56 + 40 = 166$

Since there are 24 blocks in the carton, the total surface area of these 24 blocks is $24 \times 166 = 3{,}984$. So the answer is choice (E).

You've covered the solid geometry topics that you're likely to encounter on Math 1. You've reviewed the facts and formulas and learned some useful strategies. Now it's time for you to try another set of typical solid geometry questions.

THINGS TO REMEMBER:

Surface Area of a Rectangular Solid

If the length is ℓ, the width is w, and the height is h, the formula is

Surface Area = $2\ell w + 2wh + 2\ell h$

Distance Between Opposite Vertices of a Rectangular Solid

Distance = $\sqrt{\ell^2 + w^2 + h^2}$

Volume Formulas for Uniform Solids

Volume of a Uniform Solid = Bh

Volume of a Rectangular Solid = ℓwh

Volume of a Cube = s^3

Volume of a Cylinder = $\pi r^2 h$

SOLID GEOMETRY FOLLOW-UP TEST

5 Questions (6 Minutes)

Directions: Solve the following problems. Fill in the oval corresponding to the best answer choice in the grid to the right of each question. (Answers and explanations begin on page 108.)

DO YOUR FIGURING HERE

Reference Information: Use the following formulas as needed.

Right circular cone: If r = radius and h = height, then Volume = $\frac{1}{3}\pi r^2 h$, and if c = circumference of the base and ℓ = slant height, then Lateral Area = $\frac{1}{2}c\ell$.

Sphere: If r = radius, then Volume = $\frac{4}{3}\pi r^3$ and Surface Area = $4\pi r^2$.

Pyramid: If B = area of the base and h = height, then Volume = $\frac{1}{3}Bh$.

---

1. In Figure 1, the radius of the base of the right circular cone is 3. If the volume of the cone is 12π, what is the lateral area of the cone?

    (A) 12π
    (B) 15π
    (C) 18π
    (D) 36π
    (E) 72π Ⓐ Ⓑ Ⓒ Ⓓ Ⓔ

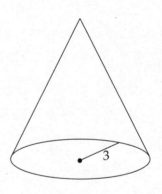

Figure 1

2. In Figure 2, d is the distance from vertex A to vertex B. What is the volume of the cube in terms of d?

   (A) $9d^3\sqrt{3}$

   (B) $3d^3\sqrt{3}$

   (C) $d^3\sqrt{3}$

   (D) $\dfrac{d^3\sqrt{3}}{3}$

   (E) $\dfrac{d^3\sqrt{3}}{9}$ Ⓐ Ⓑ Ⓒ Ⓓ Ⓔ

3. A cube with edge of length 4 is divided into 8 identical cubes. How much greater is the combined surface area of the 8 smaller cubes than the surface area of the original cube?

   (A) 48
   (B) 56
   (C) 96
   (D) 288
   (E) 384   Ⓐ Ⓑ Ⓒ Ⓓ Ⓔ

4. When a right triangle of area 3 is rotated 360° about its shorter leg, the solid that results has a volume of 30. What is the volume of the solid that results when the same right triangle is rotated about its longer leg?

   (A) 0.99
   (B) 7.90
   (C) 8.88
   (D) 31.42
   (E) 41.12   Ⓐ Ⓑ Ⓒ Ⓓ Ⓔ

5. The pyramid in Figure 3 is composed of a square base of area 12 and four equilateral triangles. What is the volume of the pyramid?

   (A) 9.8
   (B) 14.7
   (C) 17.0
   (D) 19.6
   (E) 29.4   Ⓐ Ⓑ Ⓒ Ⓓ Ⓔ

DO YOUR FIGURING HERE

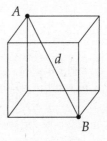

Figure 2

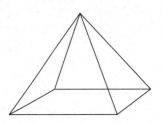

Figure 3

KAPLAN

FOLLOW-UP TEST—ANSWERS AND EXPLANATIONS

1. (B)

The formula for the lateral surface area of a cone (which will be provided in the test booklet) is in terms of c = circumference and ℓ = slant height. You can use the given base radius to get the circumference:

$$c = 2\pi r = 2\pi(3) = 6\pi$$

You can think of the slant height as the hypotenuse of a right triangle whose legs are the base radius and height of the cone.

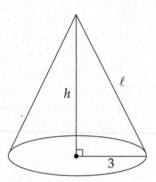

To get ℓ, first you need to find h:

$$\text{Volume of Cone} = \frac{1}{3}\pi r^2 h$$

$$12\pi = \frac{1}{3}\pi(3^2)h$$

$$12\pi = 3\pi h$$

$$h = \frac{12\pi}{3\pi} = 4$$

Now you can see that the triangle is a 3-4-5 and that $\ell = 5$. Now plug $c = 6\pi$ and $\ell = 5$ into the lateral area formula:

$$\text{Lateral Area} = \frac{1}{2}c\ell = \frac{1}{2}(6\pi)(5) = 15\pi$$

2. E

Use the formula for the distance between opposite vertices:

$$d = \sqrt{\ell^2 + w^2 + h^2}$$

What you have is a cube, so the length, width, and height are all the same—call them each x:

$$d = \sqrt{x^2 + x^2 + x^2} = \sqrt{3x^2} = x\sqrt{3}$$

$$x = \frac{d}{\sqrt{3}}$$

Now you have the length of an edge in terms of d. Cube that and you have the volume in terms of d:

$$\text{Volume} = (\text{edge})^3$$

$$= x^3$$

$$= \left(\frac{d}{\sqrt{3}}\right)^3$$

$$= \frac{d^3}{3\sqrt{3}}$$

$$= \frac{d^3}{3\sqrt{3}} \times \frac{\sqrt{3}}{\sqrt{3}}$$

$$= \frac{d^3\sqrt{3}}{9}$$

3. C

When a cube of edge length 4 is divided into 8 identical smaller cubes, the edge of each of the smaller cubes is 2:

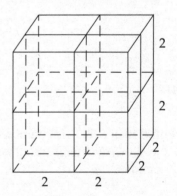

The surface area of the original cube is $6 \times 4 \times 4 = 96$. The surface area of one of the smaller cubes is $6 \times 2 \times 2 = 24$. There are eight small cubes, so their combined surface area is $8 \times 24 = 192$. The difference is $192 - 96 = 96$.

4. B

When you rotate a right triangle about a leg, you get a cone:

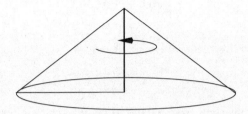

One leg becomes the base radius, and the other becomes the height of the cone. Call the long leg a and the short leg b and plug them into the cone volume formula:

$$\text{Volume of Cone} = \frac{1}{3}\pi r^2 h = \frac{1}{3}\pi a^2 b$$

It's also given that the area of the right triangle is 3, so

$$\text{Area of Right Triangle} = \frac{1}{2}\left(\text{leg}_1\right)\left(\text{leg}_2\right)$$

$$3 = \frac{1}{2}ab$$

$$ab = 6$$

Plug $ab = 6$ into the expression for the volume of the cone and you can solve for a:

$$\frac{1}{3}\pi a^2 b = 30$$

$$\frac{1}{3}\pi a(ab) = 30$$

$$\frac{1}{3}\pi a(6) = 30$$

$$2\pi a = 30$$

$$a = \frac{15}{\pi}$$

Then you can plug $a = \frac{15}{\pi}$ into the equation $ab = 6$ to solve for b:

$$ab = 6$$

$$\left(\frac{15}{\pi}\right)b = 6$$

$$b = \frac{6\pi}{15} = \frac{2\pi}{5}$$

When the triangle is rotated about its longer leg, a becomes the height and b becomes the base radius, so

$$\text{Volume} = \frac{1}{3}\pi r^2 h = \frac{1}{3}\pi b^2 a = \frac{1}{3}\pi\left(\frac{2\pi}{5}\right)^2\left(\frac{15}{\pi}\right) = \frac{60\pi^3}{75\pi}$$

$$= \frac{4}{5}\pi^2 \approx 7.90$$

5. A

To find the volume of a pyramid, you need the area of the base, which is given here as 12, and you need the height, which here you have to figure out. Imagine a triangle that includes the height and one of the lateral edges:

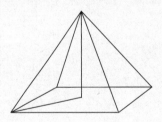

This triangle is a right triangle. The hypotenuse is the same as a side of one of the equilateral triangles, which is the same as a side of the square, which is the square root of 12, or $2\sqrt{3}$.

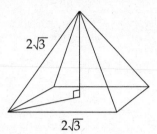

One of the legs of this right triangle is half of a diagonal of the square base—that is, half of $\left(2\sqrt{3}\right)\sqrt{2}$:

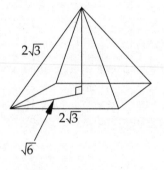

Now you can use the Pythagorean theorem to find the height:

$$\left(\sqrt{6}\right)^2 + h^2 = \left(2\sqrt{3}\right)^2$$
$$6 + h^2 = 12$$
$$h^2 = 6$$
$$h = \sqrt{6}$$

Finally you have what you need to use the volume formula:

$$\text{Volume} = \frac{1}{3}Bh = \frac{1}{3}(12)\sqrt{6} = 4\sqrt{6} \approx 9.8$$

Chapter 7: **Coordinate Geometry**

- Midpoints, distances, and slopes
- Absolute values
- Inequalities
- Circles and parabolas

A typical SAT Mathematics subject test has about five coordinate geometry questions. But that's counting just the questions that are primarily about coordinate geometry. A lot of trigonometry and functions questions assume an understanding of coordinate geometry. In fact, the material in this chapter is relevant to approximately *20 percent of Math 1 questions*.

COORDINATE GEOMETRY FACTS AND FORMULAS IN THIS CHAPTER

- Midpoint
- Distance Formula
- Slope-Intercept Form
- Slope Formula
- Positive and Negative Slopes
- Slopes of Parallel and Perpendicular Lines
- Circle
- Parabola

HOW TO USE THIS CHAPTER

Maybe you already know all the coordinate geometry you need. You can find out by taking the Coordinate Geometry Diagnostic Test. Check your answers using the answer key following the test. No matter how you score, don't worry! The answer key also shows where to find a detailed explanation for each question. The "Find Your Study Plan" section that follows the test will suggest next steps based on your performance on the Diagnostic.

Find Your Level

How you use this chapter really depends on how much time you have to prep. Find your level and pace below.

Standard Plan. Take the Coordinate Geometry Diagnostic Test. Read the rest of the chapter. Then try the Follow-Up Test at the end.

Shortcut: Take the Coordinate Geometry Diagnostic Test. The "Find Your Study Plan" section that follows the test will suggest next steps based on your Diagnostic Test score.

Panic Plan? Look through the chapter quickly and make sure you're comfortable with the material. If you're not comfortable with coordinate geometry, you should probably spend at least a little time with this chapter before moving on.

COORDINATE GEOMETRY DIAGNOSTIC TEST

5 Questions (6 Minutes)

Directions: Solve problems 1–5. Fill in the oval corresponding to the best answer choice in the grid to the right of each question. (Answers are on page 115.)

DO YOUR FIGURING HERE

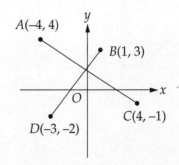

Figure 1

1. In Figure 1, what is the distance from the midpoint of segment *AC* to the midpoint of segment *BD* ?

   (A)   1.118
   (B)   1.414
   (C)   1.803
   (D)   2.236
   (E)   2.828   Ⓐ Ⓑ Ⓒ Ⓓ Ⓔ

2. If points $(6, 0)$, $(0, 0)$, $(0, 2)$, and $(a, 2)$ are consecutive vertices of a trapezoid of area 7.5, what is the value of *a* ?

   (A) 1.5   (B) 2   (C) 2.5   (D) 5   (E) 9

   Ⓐ Ⓑ Ⓒ Ⓓ Ⓔ

3. Which of the following lines has no point of intersection with the line $y = 4x + 5$?

   (A) $y = \dfrac{1}{4}x - 5$

   (B) $y = -\dfrac{1}{4}x - 5$

   (C) $y = 4x + \dfrac{1}{5}$

   (D) $y = -4x + \dfrac{1}{5}$

   (E) $y = -4x - \dfrac{1}{5}$ Ⓐ Ⓑ Ⓒ Ⓓ Ⓔ

KAPLAN)

4. Which of the following shaded regions shows the graph of the inequality $y \leq |x + 2|$?

(A)
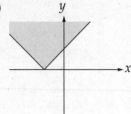

(B)
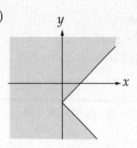

(C)
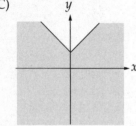

(D)
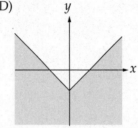

(E)
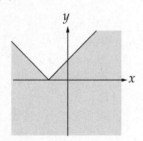

Ⓐ Ⓑ Ⓒ Ⓓ Ⓔ

5. Which of the following equations describes the set of all points (x, y) in the coordinate plane that are a distance of 5 from the point $(-3, 4)$?

    (A) $(x + 3) + (y - 4) = 5$
    (B) $(x - 3) + (y + 4) = 5$
    (C) $(x + 3)^2 + (y - 4)^2 = 5$
    (D) $(x + 3)^2 + (y - 4)^2 = 25$
    (E) $(x - 3)^2 + (y + 4)^2 = 25$

    Ⓐ Ⓑ Ⓒ Ⓓ Ⓔ

Find Your Study Plan

The answer key shows where in this chapter to find explanations for the questions you missed. Here's how you should proceed based on your Diagnostic Test score.

5: Superb! If there's not much time before test day, then you might consider skipping this chapter—you seem to know it all already! If you just want to try your hand at some more coordinate geometry questions, go to the Follow-Up Test at the end of this chapter.

3–4: Good. You have a decent grasp of coordinate geometry. But this topic is fundamental enough that you might want to read at least those parts of this chapter that relate to the questions you were unable to answer correctly.

0–2: You need to work on coordinate geometry. This topic is fundamental. Read the rest of this chapter, study the examples, and see if you can do better on the Follow-Up Test at the end of the chapter.

COORDINATE GEOMETRY TEST TOPICS

The questions in the Coordinate Geometry Diagnostic Test are typical of those on Math 1. In this chapter we'll use these questions to review the coordinate geometry you're expected to know for both levels of the SAT Mathematics. We will also use these questions to demonstrate some effective problem-solving techniques, alternative methods, and test-taking strategies that apply to SAT subject tests coordinate geometry questions.

Midpoints and Distances

Some of the more basic SAT subject tests' coordinate geometry questions are ones that concern themselves with the layout of the grid, the location of points, distances between them, midpoints, and so on. Example 1 involves both distances and midpoints.

Example 1

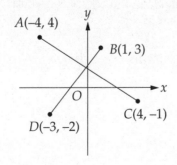

In Figure 1, what is the distance from the midpoint of segment AC to the midpoint of segment BD ?

(A) 1.118   (B) 1.414   (C) 1.803   (D) 2.236   (E) 2.828

> ### MIDPOINT
>
> To find the midpoint between (x_1, y_1) and (x_2, y_2), average the x-coordinates and average the y-coordinates:
>
> $$\text{Midpoint} = \left(\frac{x_1 + x_2}{2}, \frac{y_1 + y_2}{2} \right)$$

To find the midpoint of a segment, average the x-coordinates and average the y-coordinates of the endpoints:

$$\text{midpoint of } \overline{AC} = \left(\frac{-4+4}{2}, \frac{4-1}{2} \right) = (0, 1.5)$$

$$\text{midpoint of } \overline{BD} = \left(\frac{-3+1}{2}, \frac{-2+3}{2} \right) = (-1, 0.5)$$

To find the distance between two points, use the distance formula:

$$\text{Distance} = \sqrt{(x_2 - x_1)^2 + (y_2 - y_1)^2}$$

> ### DISTANCE FORMULA
>
> To find the distance between (x_1, y_1) and (x_2, y_2):
>
> Distance =
>
> $$\sqrt{(x_2 - x_1)^2 + (y_2 - y_1)^2}$$

The distance from (0, 1.5) to (–1, 0.5) is

$$\text{Distance} = \sqrt{(-1-0)^2 + (0.5-1.5)^2}$$
$$= \sqrt{1+1}$$
$$= \sqrt{2}$$

You could use your calculator to get a decimal approximation of $\sqrt{2}$, but you should be able to spot it as (B) 1.414. The answer is (B).

Geometry on the Grid

Finding midpoints and distances is basically a geometry thing. Here's another example of a coordinate geometry question that's essentially plane geometry transferred to the coordinate plane.

Example 2

If points (6, 0), (0, 0), (0, 2), and (a, 2) are consecutive vertices of a trapezoid of area 7.5, what is the value of a ?

(A) 1.5   (B) 2   (C) 2.5   (D) 5   (E) 9

Sketch a diagram to help comprehend the situation. Plot the three given points and connect them to make two of the sides of the trapezoid:

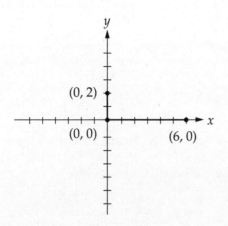

The fourth vertex has a y-coordinate of 2, so it must be somewhere along the line $y = 2$:

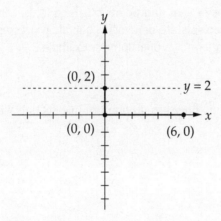

That will make the top and bottom sides of the trapezoid the parallel bases. The formula for the area of a trapezoid is

$$\text{Area of Trapezoid} = \left(\frac{\text{base}_1 + \text{base}_2}{2} \right) \times \text{height}$$

Here it's given that the area is 7.5, and you can see from the figure that one base is 6 and the height is 2. That's enough to solve for the other base.

$$7.5 = \left(\frac{6 + \text{base}_2}{\cancel{2}} \right) \times \cancel{2}$$

$$\text{base}_2 = 7.5 - 6 = 1.5$$

The top base is 1.5, and for the four vertices to be consecutive, the coordinates of the fourth vertex are (1.5, 2).

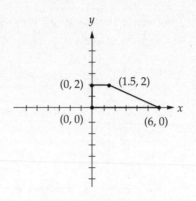

Figure 1

So $a = 1.5$, and the answer is (A).

Slope-Intercept Form

With the topic of slope-intercept form, we move from coordinate geometry that is primarily geometric into coordinate geometry that is primarily algebraic. Slopes and intercepts are descriptions of lines and points on the grid, but the processes of finding and using slopes and intercepts are generally algebraic processes. There's no need, for instance, to sketch a diagram for a question like Example 3.

Example 3

Which of the following lines has no point of intersection with the line $y = 4x + 5$?

(A) $y = \dfrac{1}{4}x - 5$

(B) $y = -\dfrac{1}{4}x - 5$

(C) $y = 4x + \dfrac{1}{5}$

(D) $y = -4x + \dfrac{1}{5}$

(E) $y = -4x - \dfrac{1}{5}$

What does *has no intersection with* mean? It means that the lines are parallel, which in turn means that the lines have the same slope. If you know the slope-intercept form, you're able to spot the correct answer instantly.

When an equation is in the form $y = mx + b$, the letter m represents the slope, and the letter b represents the y-intercept. The equation in the stem is $y = 4x + 5$. That's in slope-intercept form, so the coefficient of x is the slope.

$$y = \textcircled{4}x + 5$$

$$\text{slope} = 4$$

Now look for the answer choice with the same slope. Conveniently, all the answer choices are presented in slope-intercept form, so spotting the one with $m = 4$ is a snap. It's (C):

$$y = \textcircled{4}x + \frac{1}{5}$$

$$\text{slope} = 4$$

People who are good at memorizing methods and formulas are not necessarily the ones who get the best scores on the SAT subject tests. People who have a deeper understanding of mathematics do. If you really want to ace coordinate geometry questions, it's not enough to memorize the midpoint formula, the distance formula, the slope definition, the slope-intercept equation form, and so on. What you want is to have a real grasp of what slope is and what perpendicular and parallel lines and positive, negative, zero, and undefined slopes tell you.

Slope is a description of the "steepness" of a line. Lines that go uphill (from left to right) have positive slopes:

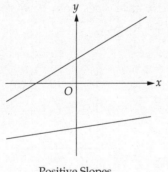

Positive Slopes

Lines that go downhill have negative slopes:

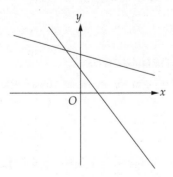

Negative Slopes

DON'T JUST MEMORIZE— INTERNALIZE

The best scores go to test takers who don't just memorize methods and formulas, but really understand the underlying math.

POSITIVE AND NEGATIVE SLOPES

Lines that go uphill from left to right have positive slopes. The steeper the uphill grade, the greater the slope.

Lines that go downhill from left to right have negative slopes. The steeper the downhill grade, the less the slope.

KAPLAN

Lines parallel to the *x*-axis have slope = 0, and lines parallel to the *y*-axis have undefined slope.

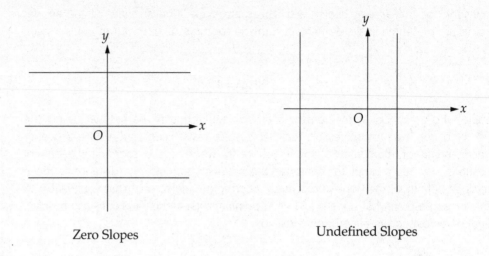

Zero Slopes   Undefined Slopes

Lines that are parallel to each other have the same slope, and lines that are perpendicular to each other have negative-reciprocal slopes.

In the figure below, the two parallel lines both have slope = 2, and the line that's perpendicular to them has slope $= -\frac{1}{2}$:

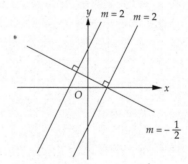

Parallel and Perpendicular Lines

Absolute Value and Inequalities

Among the more mind-bending coordinate geometry questions you'll find in the Math 1 test are ones like Example 4 that entail graphing absolute value and inequalities.

Example 4

Which of the following shaded regions shows the graph of the inequality $y \leq |x + 2|$?

(A)

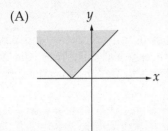

(B)

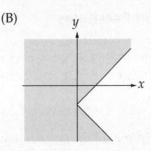

(C)

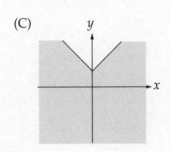

(D)

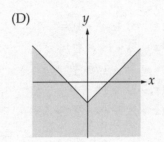

(E)

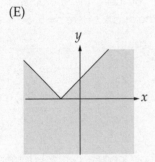

The way to handle an inequality is to think of it as an equation first, plot the line, and then figure out which side of the line to shade. This inequality is extra complicated because of the absolute value. When you graph an equation with absolute value, you generally get a line with a sharp bend, as in all of the answer choices above. To find the graph of an absolute value equation, figure out where that bend is. In this case $|x + 2|$ has a turning-point value of 0 when $x + 2 = 0$, which happens when $x = -2$. So the bend is at the point $(-2, 0)$. That narrows the choices down to (A) and (E).

Next, figure out which side gets shaded. Pick a convenient point on either side and see if that point's coordinates fit the given inequality. The point $(0, 0)$ is an easy one to work with. Do those coordinates satisfy the inequality?

$$y \le |x + 2|$$

$$0 \overset{?}{\le} |0 + 2|$$

$$0 \overset{?}{\le} 2 \quad \text{Yes.}$$

The point $(0, 0)$ must be on the shaded side of the bent line. The answer is (E).

PICK A POINT OR TWO

When you're trying to find an equation that fits a graph, pick a point or two from the graph and try them in the equations.

Circles and Parabolas

The only curved graphs you'll find on a Math 1 test are circles and parabolas. Like slope-intercept questions, these questions are essentially algebraic. Circles questions, like Example 5, are often just a matter of recalling and applying the appropriate equation.

> ### CIRCLE
>
> The equation of a circle centered at (h, k) and with radius r is
> $(x - h)^2 + (y - k)^2 = r^2$.

Example 5

Which of the following equations describes the set of all points (x, y) in the coordinate plane that are a distance of 5 from the point $(-3, 4)$?

(A) $(x + 3) + (y - 4) = 5$

(B) $(x - 3) + (y + 4) = 5$

(C) $(x + 3)^2 + (y - 4)^2 = 5$

(D) $(x + 3)^2 + (y - 4)^2 = 25$

(E) $(x - 3)^2 + (y + 4)^2 = 25$

To use the formula for the equation of a circle, you need the coordinates (h, k) of the center. Here they're given: $(-3, 4)$. And you need the radius r. Here $r = 5$. So the equation is

$$(x - h)^2 + (y - k)^2 = r^2$$
$$(x + 3)^2 + (y - 4)^2 = 25$$

The answer is (D).

There are all kinds of parabolas, and there's no simple, general parabola formula for you to memorize. You should know, however, that the graph of the general quadratic equation $y = ax^2 + bx + c$ is a parabola. It's one that opens either on top or on bottom and has an axis of symmetry parallel to the y-axis. Here, for example are parabolas representing the equations $y = x^2 - 2x + 1$ (dotted) and $y = -x^2 + 4$ (straight).

> ### PARABOLA
>
> The graph of the general quadratic equation $y = ax^2 + bx + c$ is a parabola with an axis of symmetry parallel to the y-axis.

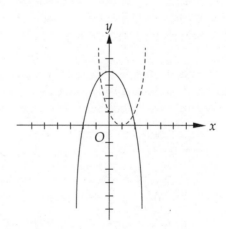

THINGS TO REMEMBER:

$$\text{Midpoint} = \left(\frac{x_1 + x_2}{2}, \frac{y_1 + y_2}{2} \right)$$

$$\text{Distance} = \sqrt{\left(x_2 - x_1\right)^2 + \left(y_2 - y_1\right)^2}$$

Slope-Intercept Form

For any equation in the form $y = mx + b$,
m = slope
b = y-intercept

Slope Formula

$$\text{Slope} = \frac{y_2 - y_1}{x_2 - x_1}$$

Slopes of Parallel and Perpendicular Lines

Parallel lines have the same slope.
Perpendicular lines have negative-reciprocal slopes.

Circle

The equation of a circle centered at (h, k) and with radius r is

$$(x - h)^2 + (y - k)^2 = r^2$$

Parabola

The graph of the general quadratic equation $y = ax^2 + bx + c$ is a parabola with an axis of symmetry parallel to the y-axis.

COORDINATE GEOMETRY FOLLOW-UP TEST

5 Questions (6 Minutes)

Directions: Solve problems 1–5. Fill in the oval corresponding to the best answer choice in the grid to the right of each question. (Answers and explanations begin on page 126.)

DO YOUR FIGURING HERE

1. The graph of the equation $x^2 + y^2 = 25$ includes how many points (x, y) in the coordinate plane where x and y are both integers?

    (A) 4   (B) 5   (C) 8   (D) 10   (E) 12

    Ⓐ Ⓑ Ⓒ Ⓓ Ⓔ

2. In Figure 1, if the midpoints of segments AB, CD, and EF are connected, what is the area of the resulting triangle?

    (A) 1   (B) 1.5   (C) 2   (D) 2.5   (E) 3

    Ⓐ Ⓑ Ⓒ Ⓓ Ⓔ

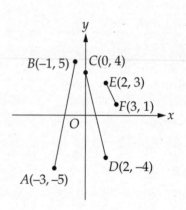

Figure 1

3. If the line $y = 8x - 24$ intersects the line $y = mx + 12$ in the fourth quadrant, which of the following must be true?

    (A) $m < -4$
    (B) $-4 < m < 3$
    (C) $-3 < m < 3$
    (D) $0 < m < 4$
    (E) $m > 3$

    Ⓐ Ⓑ Ⓒ Ⓓ Ⓔ

4. Which of the following lines is perpendicular to
    $y = -2x + 3$ and has the same y-intercept as $y = 2x - 3$?

    (A) $y = \dfrac{1}{2}x + 3$

    (B) $y = \dfrac{1}{2}x - 3$

    (C) $y = -\dfrac{1}{2}x + 3$

    (D) $y = 2x + 3$

    (E) $y = 2x - 3$ Ⓐ Ⓑ Ⓒ Ⓓ Ⓔ

5. The shaded portion of Figure 2 shows the graph of which of
    the following?

    (A) $x(y - 2x) \geq 0$

    (B) $x(y - 2x) \leq 0$

    (C) $x\left(y + \dfrac{1}{2}x\right) \geq 0$

    (D) $x\left(y - \dfrac{1}{2}x\right) \leq 0$

    (E) $x\left(y + \dfrac{1}{2}x\right) \leq 0$ Ⓐ Ⓑ Ⓒ Ⓓ Ⓔ

DO YOUR FIGURING HERE

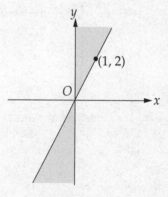

Figure 2

FOLLOW-UP TEST—ANSWERS AND EXPLANATIONS

1. E

The graph of $x^2 + y^2 = 25$ is a circle centered at the origin with a radius of 5. The square of that radius must equal the sum of the squares of the coordinates of any point on the circle. This happens on the axes.

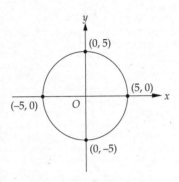

And x and y are both integers at these points as well.

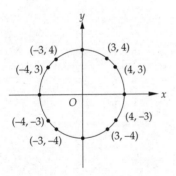

That's a total of 12. Remember, keep your eyes peeled for 3-4-5 triangles!

2. E

First find the three midpoints.

$$\text{midpoint of } \overline{AB} = \left(\frac{-3+(-1)}{2}, \frac{-5+5}{2} \right) = (-2, 0)$$

$$\text{midpoint of } \overline{CD} = \left(\frac{0+2}{2}, \frac{4+(-4)}{2} \right) = (1, 0)$$

$$\text{midpoint of } \overline{EF} = \left(\frac{2+3}{2}, \frac{3+1}{2} \right) = (2.5, 2)$$

So the triangle looks like this.

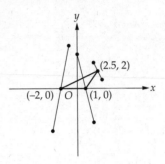

The base is 3 and the height is 2, so the area is $\frac{1}{2}$ (3)(2) = 3.

3. A

The y-intercept of the line $y = 8x - 24$ is −24. Find the x-intercept by plugging in $y = 0$:

$$0 = 8x - 24$$
$$8x = 24$$
$$x = 3$$

The y-intercept of the line $y = mx + 12$ is +12. To intersect in the fourth quadrant—which is the lower right quadrant—this second line has to have a negative slope. Not just any negative slope, but one negative enough to get it down across the x-axis before it hits the other line. In other words, the x-intercept of this second line has to be somewhere between the origin and the point (3, 0).

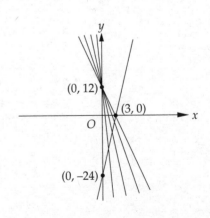

A line that goes from (0, 12) to (3, 0) would have a slope of $\dfrac{0-12}{3-0} = -4$, so the slope of the line $y = mx + 12$ must be less than -4.

4.   B

A line that's perpendicular to $y = -2x + 3$ has a slope that's the negative reciprocal of -2, which is $\dfrac{1}{2}$. That narrows the choices to (A) and (B). The y-intercept of $y = -2x - 3$ is -3, and that's the y-intercept in (B).

5.   A

Each of the answer choices is in the form of the product of two factors on the left and a "≥ 0" or "≤ 0" on the right. The product will be negative when the two factors have opposite signs, and it will be positive when the factors have the same sign. Choice (A), for example, has a "≥ 0," so you'll be looking for the factors to have the same sign.

Either:

$$x \geq 0 \text{ and } y - 2x \geq 0$$
$$x \geq 0 \text{ and } \qquad y \geq 2x$$

Or:

$$x \leq 0 \text{ and } y - 2x \leq 0$$
$$x \leq 0 \text{ and } \qquad y \leq 2x$$

The graph of $x \geq 0$ and $y \geq 2x$ looks like this:

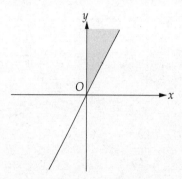

The graph of $x \leq 0$ and $y \leq 2x$ looks like this:

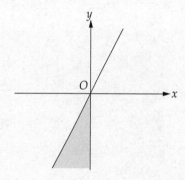

Together they make the graph in the figure.

Chapter 8: **Trigonometry**

- SOHCAHTOA
- Identities

Trigonometry is a relatively small topic on Math 1. A typical test includes only about three trigonometry questions, which are generally among the more difficult questions in the latter half of the test. Bottom line: You could ignore the trigonometry questions and still get a good score on Math 1.

> **TRIGONOMETRY FACTS AND FORMULAS IN THIS CHAPTER**
>
> - SOHCAHTOA
> - Level 1 Identities

HOW TO USE THIS CHAPTER

Maybe you already know all the trigonometry you need. You can find out by taking the Trigonometry Diagnostic Test. Check your answers using the answer key following the test. No matter how you score, don't worry! The answer key also shows where to find a detailed explanation for each question. The "Find Your Study Plan" section that follows the test will suggest next steps based on your performance on the Diagnostic.

Find Your Level

How you use this chapter really depends on how much time you have to prep. Find your level and pace below.

Standard Plan. Try the Trigonometry Diagnostic Test. Read the rest of the chapter. Then try the Follow-Up Test at the end of the chapter.

> *Shortcut: Try the Trigonometry Diagnostic Test. If you can answer the questions correctly, then you should probably skip ahead to the next chapter. If you're not too pressed for time, you could also try the questions in the Follow-Up Test at the end of the chapter.*

Panic Plan. Skip this chapter.

TRIGONOMETRY DIAGNOSTIC TEST

2 Questions (3 Minutes)

Directions: Solve problems 1–2. Fill in the oval corresponding to the best answer choice in the grid to the right of each question. (Answers are on page 131.)

DO YOUR FIGURING HERE

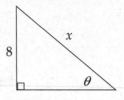

Figure 1

1. In the right triangle in Figure 1, if $\theta = 39°$, what is the value of x ?

    (A) 9.9   (B) 10.3   (C) 11.3   (D) 12.7   (E) 13.9

    Ⓐ Ⓑ Ⓒ Ⓓ Ⓔ

2. $(2 \sin x)(3 \sin x) + (6 \cos x)(\cos x) =$

    (A)   1
    (B)   6
    (C)   12
    (D) $5 \sin x + 7 \cos x$
    (E) $6 \sin x + 6 \cos x$ Ⓐ Ⓑ Ⓒ Ⓓ Ⓔ

Find Your Study Plan

The answer key shows where in this chapter to find explanations for the questions you missed. Here's how you should proceed based on your Diagnostic Test score.

2: Superb! You're already good enough with trigonometry for Math 1. If you're on the Shortcut Plan, you might consider skipping this chapter. Or, if you want, you could just go straight to the Follow-Up Test at the end of the chapter.

1: Good. You're probably already good enough with trigonometry for the Math 1 test. If you're on the Shortcut Plan, you might consider skipping this chapter. But you should at least look at the part of this chapter that discusses the question you did not get right. Then, if you want, you could just go straight to the Follow-Up Test at the end of the chapter.

0: Trigonometry may be a problem area for you, so you'd better spend some time with this chapter unless you're on the Panic Plan.

> ### DON'T BE FOOLED BY DISGUISES
>
> Sometimes what looks like a trigonometry question turns out to be primarily an algebra question.

TRIGONOMETRY TEST TOPICS

For Math 1, there's not a whole lot you have to remember about trigonometry. All you really need to know about is the sine, cosine, and tangent of acute angles measured in degrees and a few basic trigonometric identities. We'll use the questions from the Diagnostic Test to review these basic topics.

Right Triangles and SOHCAHTOA

To remember the definitions of sine, cosine, and tangent as they apply to right triangles, use the mnemonic SOHCAHTOA. That and a calculator are all you need to answer Example 1.

> ### DIAGNOSTIC TEST ANSWER KEY
>
> **1. D**
> See "Right Triangles and SOHCAHTOA."
>
> **2. B**
> See "Level 1 Identities."

Example 1

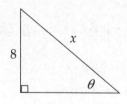

Figure 1

In the right triangle in Figure 1, if $\theta = 39°$, what is the value of x ?

(A) 9.9   (B) 10.3   (C) 11.3   (D) 12.7   (E) 13.9

The 39° angle is opposite the given 8, and the side you're looking for is the hypotenuse, so you can use the sine to find x.

$$\text{sine} = \frac{\text{opposite}}{\text{hypotenuse}}$$

$$\sin 39° = \frac{8}{x}$$

$$x = \frac{8}{\sin 39°}$$

<table>
<tr><td>

SOHCAHTOA

$\text{Sine} = \dfrac{\text{Opposite}}{\text{Hypotenuse}}$

$\text{Cosine} = \dfrac{\text{Adjacent}}{\text{Hypotenuse}}$

$\text{Tangent} = \dfrac{\text{Opposite}}{\text{Adjacent}}$

</td></tr>
</table>

Now, nobody expects you to know the sine of 39° off the top of your head. This is one of those places where you have to use your calculator. Punch in "8 ÷ sin 39 =," making sure your calculator is set to degree mode, and you'll get something like: 12.712125825, which is close to (D) 12.7.

Level 1 Identities

On the Level 1 test, you may come across questions like Example 2.

Example 2

$$(2\sin x)(3\sin x) + (6\cos x)(\cos x) =$$

(A)   1
(B)   6
(C)   12
(D) $5\sin x + 7\cos x$
(E) $6\sin x + 6\cos x$

<table>
<tr><td>

MATH 1 STRATEGY: SET YOUR CALCULATOR TO DEGREE MODE AND KEEP IT THERE

All angle measures on the Level 1 test are in degrees, so make sure your calculator is in degree mode.

</td></tr>
</table>

Start by multiplying to get rid of the parentheses and see where that will lead you:

$$(2\sin x)(3\sin x) + (6\cos x)(\cos x) = 6\sin^2 x + 6\cos^2 x$$

When you see a $\sin^2$, a plus sign, and a $\cos^2$, a little bell should ring in your head. Remember this key trigonometric identity:

$$\sin^2 x + \cos^2 x = 1$$

For any angle x, take the sine and cosine, square them both, and the squares will add up to 1. (This relationship is really just a variation of the Pythagorean theorem.) The test makers love this identity—it turns up a lot. So keep your eye out for it.

Here in Example 2, when you spot "6 sin² *x* + 6 cos² *x*," you should think immediately about how to extract "sin² *x* + cos² *x*" from it. You do that by factoring out a 6.

$$6\sin^2 x + 6\cos^2 x = 6(\sin^2 x + \cos^2 x)$$
$$= 6(1)$$
$$= 6$$

The answer is (B).

THINGS TO REMEMBER:

SOHCAHTOA

$$\text{Sine} = \frac{\text{Opposite}}{\text{Hypotenuse}}$$

$$\text{Cosine} = \frac{\text{Adjacent}}{\text{Hypotenuse}}$$

$$\text{Tangent} = \frac{\text{Opposite}}{\text{Adjacent}}$$

Pythagorean Identity

$$\sin^2 x + \cos^2 x = 1$$

TRIGONOMETRY FOLLOW-UP TEST

2 Questions (3 Minutes)

Directions: Solve problems 1–2. Fill in the oval corresponding to the best answer choice in the grid to the right of each question. (Answers and explanations begin on page 136.)

DO YOUR FIGURING HERE

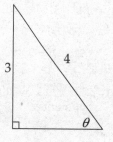

Figure 1

1. In Figure 1, what is the value of $\cos \theta$?

   (A)   0.25
   (B)   0.66
   (C)   0.75
   (D)   0.80
   (E)   1.25   Ⓐ Ⓑ Ⓒ Ⓓ Ⓔ

2. Where defined, $\dfrac{\cos^4 \theta - \sin^4 \theta}{\cos^2 \theta - \sin^2 \theta} =$

   (A)   1
   (B)   2
   (C) $\cos \theta + \sin \theta$
   (D) $\cos \theta - \sin \theta$
   (E) $\cos 2\theta + \sin 2\theta$ Ⓐ Ⓑ Ⓒ Ⓓ Ⓔ

FOLLOW-UP TEST—ANSWERS AND EXPLANATIONS

1. B

Cosine is adjacent over hypotenuse. You'll have to use the Pythagorean theorem to find the adjacent leg:

$$\text{leg} = \sqrt{4^2 - 3^2}$$
$$= \sqrt{16 - 9}$$
$$= \sqrt{7}$$
$$\cos\theta = \frac{\text{adjacent}}{\text{hypotenuse}} = \frac{\sqrt{7}}{4} \approx 0.66$$

2. A

The numerator is the difference of squares and so can be factored:

$$\frac{\cos^4\theta - \sin^4\theta}{\cos^2\theta - \sin^2\theta}$$
$$= \frac{(\cos^2\theta + \sin^2\theta)(\cos^2\theta - \sin^2\theta)}{\cos^2\theta - \sin^2\theta}$$
$$= \cos^2\theta + \sin^2\theta$$

The "$\cos^2\theta + \sin^2\theta$" should look familiar—it equals 1.

Chapter 9: **Functions**

- Substitution
- Compound functions
- Minimums and maximums
- Graphing functions

Functions is a fairly large category on the Math 1 test, which usually includes about seven functions questions. You need to be comfortable with the material in this chapter.

FUNCTIONS FACTS AND FORMULAS IN THIS CHAPTER

- Functions: Domain and Range
- Compound Functions
- Maximums and Minimums

HOW TO USE THIS CHAPTER

Maybe you already know everything you need to know about functions. You can find out by taking the Functions Diagnostic Test. The questions on the Diagnostic Test are typical of what you could expect. Check your answers using the answer key following the test. No matter how you score, don't worry! The answer key also shows where to find a detailed explanation for each question. The "Find Your Study Plan" section that follows the test will suggest next steps based on your performance on the Diagnostic.

Find Your Level

How you use this chapter really depends on how much time you have to prep. Find your level and pace below.

Standard Plan. Try the Functions Diagnostic Test. Read the rest of the chapter. Then try the Follow-Up Test at the end of the chapter.

> *Shortcut: Try the Functions Diagnostic Test and check your answers. The "Find Your Study Plan" section that follows the test will suggest next steps based on your Diagnostic Test score.*

Panic Plan. Look through the chapter quickly and make sure you're comfortable with the material. If you're not comfortable with functions, spend some time with this critical chapter.

FUNCTIONS DIAGNOSTIC TEST

4 Questions (6 Minutes)

Directions: Solve problems 1–4. Fill in the oval corresponding to the best answer choice in the grid to the right of each question. (Answers are on page 140.)

DO YOUR FIGURING HERE

1. If $f(x) = x^2 + \dfrac{x}{2}$, then $f(a + 2) =$

    (A) $a^2 + \dfrac{a}{2}$

    (B) $a^2 + \dfrac{5a}{2} + 2$

    (C) $a^2 + \dfrac{5a}{2} + 5$

    (D) $a^2 + \dfrac{9a}{2} + 2$

    (E) $a^2 + \dfrac{9a}{2} + 5$ Ⓐ Ⓑ Ⓒ Ⓓ Ⓔ

2. If $f(x) = \sqrt{x}$ and $g(x) = \sqrt{x^2 + 4}$, what is the value of $f(g(2))$?

    (A) 0   (B) 1.41   (C) 1.68   (D) 2.45   (E) 2.83

    Ⓐ Ⓑ Ⓒ Ⓓ Ⓔ

3.   What is the maximum value of $f(x) = 2 - (x + 2)^2$?

(A) −4   (B) −2   (C) 0   (D) 2   (E) 4

Ⓐ Ⓑ Ⓒ Ⓓ Ⓔ

4.   If $f(x) = |1 - x|$, which of the following could be the graph of $y = f(x)$?

(A)   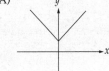   (B)   (C)

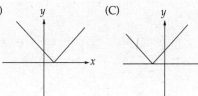

(D)   (E)

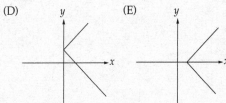

Ⓐ Ⓑ Ⓒ Ⓓ Ⓔ

Find Your Study Plan

The answer key shows where in this chapter to find explanations for the questions you missed. Here's how you should proceed based on your Diagnostic Test score.

4: Superb! You're already good enough with functions for the Math 1 test. If you're taking a "shortcut," you might consider skipping this chapter. Or, if you want, you could just go straight to the Follow-Up Test at the end of the chapter.

3: Good. You're probably already good enough with functions. If you're taking a "shortcut," you might consider skipping this chapter. But you should at least look at the part of this chapter that discusses the question you did not get right. Then, if you have time, you could go straight to the Follow-Up Test at the end of the chapter.

0–2: Functions may be a problem area for you, so you'd better spend some time with this chapter.

FUNCTIONS TEST TOPICS

We'll use the questions on the Functions Diagnostic Test to illustrate the conventions of functions.

Who's Afraid of Functions?

Lots of students are afraid of functions. But there's nothing especially difficult about them. They just look scary. Once you "get" the conventions, however, you'll never be afraid of functions again. Here's a quick and painless review of the basic things you need to know about functions.

A function is a process that turns a number into another number. Squaring is an example of a function. For any number you can think of, there is a unique number that is its square. The conventional way of writing this function is

$$f(x) = x^2$$

When you apply the function to some particular number, such as −5, you write it this way:

$$f(-5)$$

And to find the value of $f(-5)$, you plug $x = -5$ into the definition:

$$f(x) = x^2$$
$$f(-5) = (-5)^2 = 25$$

Substitution

The most straightforward functions questions you'll encounter on the Math 1 test are questions like Example 1 that ask you simply to apply a function to some number or expression.

Example 1

If $f(x) = x^2 + \dfrac{x}{2}$, then $f(a + 2) =$

(A) $a^2 + \dfrac{a}{2}$

(B) $a^2 + \dfrac{5a}{2} + 2$

(C) $a^2 + \dfrac{5a}{2} + 5$

(D) $a^2 + \dfrac{9a}{2} + 2$

(E) $a^2 + \dfrac{9a}{2} + 5$

To find $f(a + 2)$, plug $x = a + 2$ into the definition. Substituting an algebraic expression for x is a little more complicated than substituting a number for x, but the idea's the same.

$$f(x) = x^2 + \frac{x}{2}$$

$$f(a+2) = (a+2)^2 + \frac{a+2}{2}$$

$$= a^2 + 4a + 4 + \frac{a}{2} + 1$$

$$= a^2 + \frac{9a}{2} + 5$$

The answer is (E).

Compound Functions

The letter f is not the only letter used to designate a function, though it's the most popular. Second in popularity is the letter g, which is generally used in a question that includes two different functions.

Take a look at Example 2:

If $f(x)=\sqrt{x}$ and $g(x)=\sqrt{x^2+4}$, what is the value of $f(g(2))$?

(A) 0   (B) 1.41   (C) 1.68   (D) 2.45   (E) 2.83

> ## COMPOUND FUNCTIONS
>
> $f(g(x))$ means apply g first and then apply f to the result.
>
> $g(f(x))$ means apply f first and then apply g to the result.
>
> $f(x)g(x)$ means apply f and g separately and then multiply the results.

When one function is written inside another function's parentheses, apply the inside function first:

$$g(x)=\sqrt{x^2+4}$$
$$g(2)=\sqrt{2^2+4}=\sqrt{8}$$

Then apply the outside function to the result:

$$f(x)=\sqrt{x}$$
$$f\left(\sqrt{8}\right)=\sqrt{\sqrt{8}}\approx 1.68$$

The answer is (C).

Notice that there's a difference between $f(g(2))$ and $g(f(2))$. In the latter, you're to apply the function f first and then the function g to the result. The order makes a difference: $f(g(2))\approx 1.68$, but $g(f(2))\approx 2.45$ (which is, of course, included in the answer choices as a distractor).

Notice also that there's a difference between $f(g(x))$ and $f(x)g(x)$. In the former expression, a hierarchy is indicated: The function g is inside the parentheses of function f, so you apply g first, and then you apply f to the result. In the expression $f(x)g(x)$, however, the two functions are written side by side, which indicates multiplication. You apply the function f to 2, and you apply the function g to 2, and then you multiply the results:

> ## WATCH THOSE PARENTHESES
>
> When it comes to functions, pay close attention to order and parentheses.

$$f(2)=\sqrt{2}$$
$$g(2)=\sqrt{8}$$
$$f(2)g(2)=\left(\sqrt{2}\right)\left(\sqrt{8}\right)=\sqrt{16}=4$$

So when it comes to functions, be sure to pay close attention to order and parentheses.

Maximums and Minimums

Another typical functions question is one that asks for a minimum or, as in Example 3, maximum value of a function.

Example 3

What is the maximum value of $f(x) = 2 - (x + 2)^2$?

(A) −4   (B) −2   (C) 0   (D) 2   (E) 4

If you have a graphing calculator (and know how to use it), you could graph the function and trace the graph to find the maximum. But it's really a lot easier if you conceptualize the situation. The expression $2 - (x + 2)^2$ will be at its maximum when the part being subtracted from the 2 is as small as it can be. The part after the minus sign, $(x + 2)^2$, is the square of something, so it must be positive and can be no smaller than 0. When $x = -2$, $(x + 2)^2 = 0$, and the whole expression $2 - (x + 2)^2 = 2 - 0 = 2$. For any other value of x, the part after the minus sign will be greater than 0, and the whole expression will be less than 2. So 2 is the maximum value, and the answer is (D).

> **MAXIMUMS AND MINIMUMS**
>
> To find a maximum or minimum value of a function, look for parts of the expression—especially squares—that have upper or lower limits.

Graphing Functions

Like almost everything else with functions, graphing is no big deal once you understand the conventions. Example 4 provides a very good illustration.

Example 4

If $f(x) = |1 - x|$, which of the following could be the graph of $y = f(x)$?

(A)
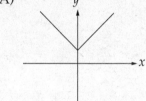

(B)
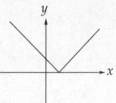

(C)
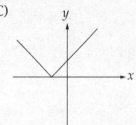

(D)
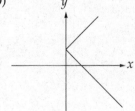

(E)
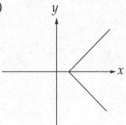

The question says $f(x) = |1 - x|$ and asks for the graph of $y = f(x)$. Well, that just means the graph of $y = |1 - x|$.

The smallest that y can be is 0, because the absolute value of anything is nonnegative. That narrows the choices to (B) and (C). Notice that (D) and (E) cannot be functions, because for some values of x, they show two associated values of y. A function, by definition, yields no more than one y for any particular x.

To choose between (B) and (C), think about what value of x yields y = 0.

$$|1-x|=0$$
$$1-x=0$$
$$-x=-1$$
$$x=1$$

So the graph touches the x-axis on the positive side, and the answer is (B).

Now that you've had a good look at some typical Math 1 test functions questions, it's time to try a few more on your own. See how well you can do with the questions in the Functions Follow-Up Test.

THINGS TO REMEMBER:

Functions: Domain and Range

A **function** is a set of ordered pairs (x, y) such that for each value of x, there is one and only one value of y.

The set of all allowable x values is called the **domain**. Note that, according to the Math 1 test directions, the domain of any function f is assumed to be the set of all real numbers x for which $f(x)$ is a real number.

The corresponding set of all y values is called the **range**.

Compound Functions

$f(g(x))$ means apply g first and then apply f to the result.

$g(f(x))$ means apply f first and then apply g to the result.

$f(x)g(x)$ means apply f and g separately and then multiply the results.

Maximums and Minimums

To find a maximum or minimum value of a function, look for parts of the expression—especially squares—that have upper or lower limits.

FUNCTIONS FOLLOW-UP TEST

4 Questions (6 Minutes)

Directions: Solve problems 1–4. Fill in the oval corresponding to the best answer choice in the grid to the right of each question. (Answers and explanations are on page 148.)

DO YOUR FIGURING HERE

1. If $f(x) = x^2 + x$ for all x, and if $f(a - 1) = -\dfrac{1}{4}$, what is the value of a ?

    (A) $-\dfrac{1}{2}$

    (B) $-\dfrac{1}{4}$

    (C) $\dfrac{1}{4}$

    (D) $\dfrac{1}{2}$

    (E) $\dfrac{3}{4}$ Ⓐ Ⓑ Ⓒ Ⓓ Ⓔ

2. If $f(x) = x^2$ and $g(x) = 2x$, what is the value of $f(g(-3)) - g(f(-3))$?

    (A)   54
    (B)   18
    (C)   0
    (D) –18
    (E) –54   Ⓐ Ⓑ Ⓒ Ⓓ Ⓔ

3. For what value of x is $f(x) = 7 - (4 - x)^2$ at its maximum?

    (A)   −9
    (B)   −3
    (C)   3
    (D)   4
    (E)   7   Ⓐ Ⓑ Ⓒ Ⓓ Ⓔ

4. The graph in Figure 1 could be the graph of which of the following functions?

    (A) $f(x) = (x - 2)^2$
    (B) $f(x) = x^2 - 4$
    (C) $f(x) = 4 - x^2$
    (D) $f(x) = |x^2 - 4|$
    (E) $f(x) = -|x^2 - 4|$ Ⓐ Ⓑ Ⓒ Ⓓ Ⓔ

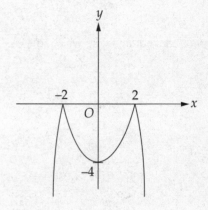

Figure 1

FOLLOW-UP TEST—ANSWERS AND EXPLANATIONS

1. D

Plug $x = a - 1$ into the definition of the function:

$$f(x) = x^2 + x$$
$$f(a-1) = (a-1)^2 + (a-1)$$
$$= a^2 - 2a + 1 + a - 1$$
$$= a^2 - a$$

Now write an equation in which the result is equal to $-\frac{1}{4}$:

$$a^2 - a = -\frac{1}{4}$$
$$a^2 - a + \frac{1}{4} = 0$$
$$4a^2 - 4a + 1 = 0$$
$$(2a-1)^2 = 0$$
$$2a - 1 = 0$$
$$2a = 1$$
$$a = \frac{1}{2}$$

2. B

Perform the inside functions first:

$$g(x) = 2x$$
$$g(-3) = 2(-3) = -6$$

$$f(x) = x^2$$
$$f(-3) = (-3)^2 = 9$$

Then perform the outside functions on the results:

$$f(x) = x^2$$
$$f(g(-3)) = f(-6) = (-6)^2 = 36$$

$$g(x) = 2x$$
$$g(f(-3)) = g(9) = 2(9) = 18$$

Now you can find the difference:

$$f(g(-3)) - g(f(-3)) = 36 - 18 = 18$$

3. D

The expression $7 - (4 - x)^2$ will be at its maximum when the part after the minus sign, $(4 - x)^2$, is as small as it can be. Something squared can never be less than zero, so the whole expression will be at its maximum when $(4 - x)^2 = 0$:

$$(4-x)^2 = 0$$
$$4 - x = 0$$
$$x = 4$$

4. E

The graph shows three points that represent solutions to the equation $y = f(x)$: $(-2, 0)$, $(2, 0)$, and $(0, -4)$. The first point satisfies choices (B), (C), (D), and (E). The second point satisfies all the answer choices. The point $(0, -4)$, however, satisfies only choices (B) and (E). How can you decide between those choices? Think about some other point on the graph. What about when $x = 3$?

Then, according to choice (B), y would be $(3)^2 - 4 = 5$ while according to (E), y would be $-|(3)^2 - 4| = -5$.

You can see from the graph that when $x = 3$, y is negative, so the only choice that fits is (E).

Chapter 10: **Miscellaneous Topics**

- Symbols, definitions, and instructions
- Imaginary and complex numbers
- Logic
- Percents, averages, and rates
- Permutations and combinations
- Probability

This chapter covers a dozen largely unrelated math topics. What these diverse topics have in common is that they are all regularly tested on the Mathematics subject tests. You will not see all the topics in this chapter on any one Math 1 test, but you can be sure that some will appear.

> **MISCELLANEOUS FACTS AND FORMULAS IN THIS CHAPTER**
>
> - Rules of Imaginary Numbers
> - Contrapositive
> - Percent Increase and Decrease Formulas
> - Average Rate Formula
> - Average, Median, and Mode
> - Permutations and Combinations Formulas
> - Probability Formulas

HOW TO USE THIS CHAPTER

The questions on the Diagnostic Test are typical of what you could expect on a Math 1 test. Check your answers using the answer key following the test. No matter how you score, don't worry! The answer key also shows where to find a detailed explanation for each question. The "Find Your Study Plan" section that follows the test will suggest next steps based on your performance on the Diagnostic.

Find Your Level

How you use this chapter really depends on how much time you have to prep. Find your level and pace below.

Standard Plan. Try the questions in the Miscellaneous Topics Diagnostic Test. Read the rest of the chapter. Then try the Follow-Up Test at the end of the chapter.

Shortcut: Try the Miscellaneous Topics Diagnostic Test. The "Find Your Study Plan" section that follows the test will suggest next steps based on your Diagnostic Test score.

Panic Plan. Make sure you're comfortable with the fundamental material in previous chapters before you spend any time on these miscellaneous topics.

MISCELLANEOUS TOPICS
DIAGNOSTIC TEST

9 Questions (11 Minutes)

Directions: Solve problems 1–9. Fill in the oval corresponding to the best answer choice in the grid to the right of each question. (Answers are on page 154.)

(Answers are on page 154.)

1. An operation § is defined for all real numbers a and b by the equation $a \ \S \ b = \dfrac{a^2}{2} + \dfrac{b}{3}$. If $-3 \ \S \ x = 0$, what is the value of x ?

   (A) -13.5 (B) -9 (C) 0 (D) 3 (E) 9

   Ⓐ Ⓑ Ⓒ Ⓓ Ⓔ

2. If $i^2 = -1$, then $(2 - 3i)^2 =$

   (A) $-5 - 12i$
   (B) $4 + 6i$
   (C) $4 + 12i$
   (D) $5 + 12i$
   (E) $13 - 6i$ Ⓐ Ⓑ Ⓒ Ⓓ Ⓔ

3. "If p, then q" is logically equivalent to which of the following?

   I. If q, then p.
   II. If not p, then not q.
   III. If not q, then not p.

   (A)   I only
   (B)   III only
   (C)   I and II only
   (D)   I and III only
   (E)   I, II, and III   Ⓐ Ⓑ Ⓒ Ⓓ Ⓔ

DO YOUR FIGURING HERE

KAPLAN

4. The charge for a taxi ride in a certain city is 2 dollars for the first one-fifth of a mile and 20 cents for each additional one-fifth of a mile. Which of the following expressions represents the charge, in cents, for a taxi ride of exactly x miles, where x is a positive integer?

    (A) $2x + 80$
    (B) $3x$
    (C) $100x + 180$
    (D) $200x + 80$
    (E) $300x$ Ⓐ Ⓑ Ⓒ Ⓓ Ⓔ

5. During the eighteenth century, the population of country X increased by 1,000 percent. During the nineteenth century, the population increased by 500 percent. By what percent did the population increase over the two-century period?

    (A)   1,500%
    (B)   1,700%
    (C)   4,900%
    (D)   5,000%
    (E)   6,500%   Ⓐ Ⓑ Ⓒ Ⓓ Ⓔ

6. George's average (arithmetic mean) score on the first four English tests of the semester is 92. If he earns a score of 77 on the fifth test, what will his new average be?

    (A) 84.5   (B) 86.0   (C) 87.5   (D) 88.0   (E) 89.0

                                      Ⓐ Ⓑ Ⓒ Ⓓ Ⓔ

7. Martha drives half the distance from A to B at 40 miles per hour and the other half of the distance at 60 miles per hour. What is her average rate of speed, in miles per hour, for the entire trip?

    (A) 48   (B) 49   (C) 50   (D) 51   (E) 52

                                      Ⓐ Ⓑ Ⓒ Ⓓ Ⓔ

8. A committee of 3 women and 2 men is to be formed from a pool of 11 women and 7 men. How many different committees can be formed?

   (A)   3,465
   (B)   6,930
   (C)   10,395
   (D)   20,790
   (E)   41,580   Ⓐ Ⓑ Ⓒ Ⓓ Ⓔ

9. The probability of rain today is 60 percent, and the independent probability of rain tomorrow is 75 percent. What is the probability that it will rain neither today nor tomorrow?

   (A) 10%   (B) 15%   (C) 45%   (D) 55%   (E) 65%

                              Ⓐ Ⓑ Ⓒ Ⓓ Ⓔ

DO YOUR FIGURING HERE

Find Your Study Plan

It doesn't matter much how you did on the Miscellaneous Topics Diagnostic Test as a whole, because the topics are so varied. What matters is exactly what questions you had trouble with. No matter what your Diagnostic Test score is, your next step is to read and study those parts of this chapter that address the particular diagnostic questions you got wrong. The answer key shows where in this chapter to find explanations for the questions you missed.

MISCELLANEOUS TEST TOPICS

We'll use the questions on the Miscellaneous Topics Diagnostic Test to point out more topics you'll need to be prepared to deal with on the SAT Subject Test: Mathematics.

Symbolism, Definitions, and Instructions

One miscellaneous question type that turns up with great regularity is the type that presents you with a new symbol representing an unfamiliar operation, or a new word with an unfamiliar definition, or a new procedure with unfamiliar steps. In Example 1 you have a new symbol and operation

Example 1

An operation § is defined for all real numbers a and b by the equation $a \S b = \dfrac{a^2}{2} + \dfrac{b}{3}$. If $-3 \S x = 0$, what is the value of x ?

(A) −13.5   (B) −9   (C) 0   (D) 3   (E) 9

The very point of a question like this is to test your ability to not freak out at the sight of something new and unfamiliar. You're not expected to have any prior experience with this operation: The test makers just made it up! But they also defined it for you, so as long as you don't freak out, everything you need to answer the question is right there.

The operation § is defined as $a \S b = \dfrac{a^2}{2} + \dfrac{b}{3}$, and when $a = -3$ and $b = x$, the result is 0.

$$a \S b = \frac{a^2}{2} + \frac{b}{3}$$

$$3 \S x = \frac{3^2}{2} + \frac{x}{3} = 0$$

$$\frac{9}{2} + \frac{x}{3} = 0$$

$$\frac{x}{3} = -\frac{9}{2}$$

$$x = -\frac{27}{2} = -13.5$$

> **DON'T FREAK OUT**
>
> If you see something—a symbol, a word—you've never seen before, chances are the test makers just made it up to test your ability to stay cool in the face of something new and unfamiliar.

The answer is (A).

Imaginary and Complex Numbers

There's a chance you will encounter a question that begins, "If $i^2 = -1$ . . . ," like the example below:

Example 2

If $i^2 = -1$, then $(2 - 3i)^2 =$

(A) $-5 - 12i$
(B) $4 + 6i$
(C) $4 + 12i$
(D) $5 + 12i$
(E) $13 - 6i$

Imaginary numbers: To answer an imaginary numbers question like this, you have to know how to use your i's. When you're adding or subtracting, i's act a lot like variables:

$$i + i = 2i$$
$$i - i = 0$$
$$2i + 3i = 5i$$
$$2i - 3i = -i$$

But when you're multiplying, or raising to a power, don't forget to take that extra step of changing i^2 to -1:

$$i \times i = i^2 = -1$$
$$(2i)(3i) = 6i^2 = 6(-1) = -6$$
$$(3i)^2 = (3i)(3i) = 9i^2 = 9(-1) = -9$$

Notice that when you raise i to successive powers, a pattern develops:

$$
\begin{array}{lll}
i^1 = i & i^5 = i & i^9 = i \\
i^2 = -1 & i^6 = -1 & i^{10} = -1 \\
i^3 = -i & i^7 = -i & i^{11} = -i \\
i^4 = 1 & i^8 = 1 & i^{12} = 1
\end{array}
$$

In Example 2, you're looking for the square of a complex number.

Example 2

To multiply complex numbers, use FOIL. (Review chapter 4 if you forgot what FOIL stands for. It's a way of remembering the order of operations for multiplying binomials: first, outer, inner, last.) Just remember again to take that extra step of changing i^2 to -1:

$$(2 - 3i)(2 - 3i)$$
$$= (2)(2) + (2)(-3i) + (-3i)(2) + (-3i)(-3i)$$
$$= 4 - 6i - 6i + 9i^2$$
$$= 4 - 12i - 9$$
$$= -5 - 12i$$

The answer is (A).

Logic

If you encounter a logic question, it is likely that about all you'll have to know is the so-called contrapositive. That's what Example 3 is getting at.

Example 3

"If p, then q" is logically equivalent to which of the following?

   I. If q, then p.
  II. If not p, then not q.
 III. If not q, then not p.

(A)   I only

(B)   III only

(C)   I and II only

(D)   I and III only

(E)   I, II, and III

RULES OF IMAGINARY NUMBERS

$ai + bi = (a + b)i$

$ai - bi = (a - b)i$

$(ai)(bi) = -ab$

$i^2 = -1$

$i^3 = -i$

$i^4 = 1$

COMPLEX NUMBERS

A number in the form $a + bi$, where a and b are real numbers, is called a complex number.

The original statement, "If *p*, then *q*," is a general form that covers such statements as the following:

1. If you live in Alabama, then you live in the United States (*p* = "live in Alabama"; *q* = "live in U.S.").
2. All prime numbers are integers (*p* = "is prime"; *q* = "is an integer").
3. If Marla studies, she will get an A on the test (*p* = "studies"; *q* = "gets an A").

Of the three Roman numeral options in Example 3, only one necessarily follows. Take a look at the options one at a time:

I. If *q*, then *p*. This is not necessarily so. You cannot simply switch the *p* and the *q*. Look what illogical results you would get with the three samples above:

1. If you live in the United States, then you live in Alabama.
2. All integers are prime numbers.
3. If Marla gets an A on the test, then she must have studied.

Can you see that none of these statements—not even number 3—necessarily follows from the original?

II. If not *p*, then not *q*. This is not necessarily so. You cannot simply negate both the *p* and the *q*. Look what illogical results you would get with the three samples above:

1. If you don't live in Alabama, then you don't live in the United States.
2. If a number's not prime, then it's not an integer.
3. If Marla doesn't study, then she won't get an A.

Can you see that none of these statements necessarily follows from the original statement?

III. If not *q*, then not *p*. This is true. If you both switch the *p* and *q* and negate them, the result is logically equivalent to the original. This is the contrapositive. Here are the contrapositives of the three samples above:

1. If you don't live in the United States, then you don't live in Alabama.
2. If a number's not an integer, then it's not a prime number.
3. If Marla doesn't get an A on the test, then she must not have studied.

These three statements are all as true as the originals they're based on. So of the three options, only III is true, and the answer is (B).

> **CONTRAPOSITIVE**
>
> "If *p*, then *q*," is logically equivalent to "If not *q*, then not *p*."

Translating from English into Algebra

Solving a word problem means taking a situation that is described verbally and turning it into one that is described mathematically. It means translating from English into algebra, which is exactly what Example 4 asks you to do.

Example 4

The charge for a taxi ride in a certain city is 2 dollars for the first one-fifth of a mile and 20 cents for each additional one-fifth of a mile. Which of the following expressions represents the charge, in cents, for a taxi ride of exactly x miles, where x is a positive integer?

(A) $2x + 80$

(B) $3x$

(C) $100x + 180$

(D) $200x + 80$

(E) $300x$

To translate successfully, you must first be sure you understand the original. It's usually a good idea to read a word problem question stem twice. The charge for the first fifth of a mile is 2 dollars. You want your answer in cents, so write:

$$\text{charge} = 200 + \cdots$$

No matter how far you go, your fare starts with 200 cents. On top of that, you pay 20 cents for each additional fifth of a mile:

$$\text{charge} = 200 + 20(\text{\# of additional fifth-miles})$$

The trickiest part of this translation is figuring out how to express the number of additional fifth-miles in terms of x.

If x is the number of whole miles, and each whole mile contains 5 fifth-miles, the total number of fifth-miles is $5x$. But you don't have to pay 20 cents for every one of those $5x$ fifth-miles—you already paid for the first fifth-mile. The number of fifth-miles after the first is $5x - 1$:

$$\text{charge} = 200 + 20(5x - 1) = 200 + 100x - 20 = 100x + 180$$

The answer is (C).

Percent Increase and Decrease

Some word problems present classic situations for which there are standard approaches, or even formulas. Example 5, for instance, requires that you know how to find a percent increase.

Example 5

During the eighteenth century, the population of country X increased by 1,000 percent. During the nineteenth century, the population increased by 500 percent. By what percent did the population increase over the two-century period?

(A) 1,500%
(B) 1,700%
(C) 4,900%
(D) 5,000%
(E) 6,500%

> **PERCENT INCREASE AND DECREASE FORMULAS**
>
> Percent increase =
>
> $\dfrac{\text{Amount of increase}}{\text{Original Amount}} \times 100\%$
>
> Percent decrease =
>
> $\dfrac{\text{Amount of decrease}}{\text{Original Amount}} \times 100\%$

First, you should realize that answering this question is going to take more than merely adding 1,000 percent and 500 percent to get (A), 1,500 percent. You're not going to find a word problem on the Math 1 test for which all you have to do is add two numbers.

This word problem is extra complicated for two reasons. First, the percents are so huge. You don't often have to work with "1,000 percent" or "500 percent." Second, you have a compound percent increase situation, a situation that's always trickier than it looks at first.

Call the population of country X at the beginning of the eighteenth century P. That number increases by 1,000 percent. That is, it goes up by 1,000 percent of itself. One thousand percent of P means 10P, so if P goes up by that much, it becomes P + 10P = 11P.

$$P + (1{,}000\% \text{ of } P) = P + 10P = 11P$$

Now the population at the beginning of the nineteenth century is 11P. That number increases by 500 percent. In other words, on top of 11P, you add 500 percent of 11P:

$$11P + (500\% \text{ of } 11P) = 11P + 5(11P) = 66P$$

The population at the end of the nineteenth century is 66P. That's 65P more than the original population P:

$$66P = P + 65P$$
$$= P + (6{,}500\% \text{ of } P)$$

The net increase is 6,500 percent, and the answer is (E).

Averages

Another classic situation with a standard approach is an averages question like Example 6.

Example 6

George's average (arithmetic mean) score on the first four English tests of the semester is 92. If he earns a score of 77 on the fifth test, what will his new average be?

(A) 84.5   (B) 86.0   (C) 87.5   (D) 88.0   (E) 89.0

The key to almost every SAT Subject Test averages question is to think about the sum. Everybody knows the standard averages formula:

$$\text{Average} = \frac{\text{Sum of the terms}}{\text{Number of terms}}$$

Perhaps less familiar, but often more useful, is this other version of the averages formula:

$$\text{Sum of the terms} = \text{Average} \times \text{Number of terms}$$

George's average after four tests is 92. To find out what effect a 77 on test number five will have, calculate the sum of the first four scores:

$$\text{Sum} = \text{Average} \times \text{Number} = 92 \times 4 = 368$$

To get the new average, add 77 and divide by 5:

$$\text{New Average} = \frac{368 + 77}{5} + \frac{445}{5} = 89.0$$

The answer is (E).

Distance, Rate, and Time

Yet another classic situation with a standard approach is the distance-rate-and-time question. Example 7 is an interesting variation.

Example 7

Martha drives half the distance from A to B at 40 miles per hour and the other half the distance at 60 miles per hour. What is her average rate of speed, in miles per hour, for the entire trip?

(A) 48   (B) 49   (C) 50   (D) 51   (E) 52

At first you might think, "Halfway at 40 mph, halfway at 60 mph—the average is 50." But that's way too simple for an SAT Subject Test: Math 1 word problem. The average rate would be 50 if Martha drove half the *time* at 40 and half the *time* at 60. But it's half the *distance* she drives at 40 and half the *distance* she drives at 60. So she actually spends more time driving at the slower speed, and her average speed will be something closer to 40 than to 60. So the answer's going to be either (A) or (B).

Average speed is defined as **total distance divided by total time:**

$$\text{Average speed} = \frac{\text{Total distance}}{\text{Total time}}$$

AVERAGE RATE FORMULA

Average A per B

$= \dfrac{\text{Total } A}{\text{Total } B}$

Average speed

$= \dfrac{\text{Total Distance}}{\text{Total Time}}$

In Example 7 the distance is not specified, so call it *d*.

To get the total time, think about the time for each half-distance. Martha travels the first $\frac{d}{2}$ miles at 40 miles per hour:

$$\text{Time}_1 = \frac{\text{Distance}}{\text{Rate}} = \frac{\frac{d}{2}}{40} = \frac{d}{80}$$

She travels the second $\frac{d}{2}$ miles at 60 miles per hour:

$$\text{Time}_2 = \frac{\text{Distance}}{\text{Rate}} = \frac{\frac{d}{2}}{60} = \frac{d}{120}$$

The total time, then, is

$$\text{Total time} = \frac{d}{80} + \frac{d}{120}$$
$$= \frac{3d}{240} + \frac{2d}{240} = \frac{5d}{240} = \frac{d}{48}$$

AVOID THE SPEED TRAP

When a problem asks for average rate or average speed, don't just average the given rates or speeds. It's not that simple. Use the Average Rate formula.

Now plug Total distance = *d* and Total time = $\frac{d}{48}$ into the general average speed formula:

$$\text{Average speed} = \frac{\text{Total distance}}{\text{Total time}} = \frac{d}{\frac{d}{48}} = 48$$

The answer is (A).

Permutations and Combinations

To be successful with combinations and permutations questions like Example 8, you have to remember the relevant formulas.

Example 8

A committee of 3 women and 2 men is to be formed from a pool of 11 women and 7 men. How many different committees can be formed?

(A) 3,465   (B) 6,930   (C) 10,395   (D) 20,790   (E) 41,580

This question asks about the number of groups that can be formed, so it's a combinations question. The number of combinations of 11 women taken 3 at a time is

$$11C_3 = \frac{11!}{3!(11-3)!}$$

$$= \frac{11!}{3!(8!)}$$

$$= \frac{11 \times 10 \times 9}{3 \times 2}$$

$$= 165$$

The number of combinations of 7 men taken 2 at a time is

$$7C_2 = \frac{7!}{2!(7-2)!}$$

$$= \frac{7!}{2!(5!)}$$

$$= \frac{7 \times 6}{2}$$

$$= 21$$

For each of the 165 combinations of women, there are 21 combinations of men, so the combined number of combinations is 165 × 21 = 3,465. The answer is (A).

If you're good at memorizing formulas—if you can master the permutations and combinations formulas without too much trouble—go ahead and do it. But all these formulas probably won't get you more than one right answer on the Math 1 test, so they don't deserve a whole lot of time and effort. Remember that most calculators will compute permutations and combinations for you.

Probability

To answer a probability question, you need to know not only how to use the general probability formula, but also how to deal with multiple probabilities and probabilities that events will not occur. You have a little bit of everything in Example 9.

Example 9

The probability of rain today is 60 percent, and the independent probability of rain tomorrow is 75 percent. What is the percent probability that it will rain neither today nor tomorrow?

(A) 10%   (B) 15%   (C) 45%   (D) 55%   (E) 65%

To get the probability that an event will not occur, subtract the probability that it will occur from 1. The probability that it will rain today is 60%, so the probability that it will not rain today is 100% – 60% = 40%. The probability that it will rain tomorrow is 75%, so the probability that it will not rain tomorrow is 100% – 75% = 25%.

To get the probability that two independent events will both occur, multiply the individual probabilities. Thus, to get the probability that it will not rain today or tomorrow, multiply 40% and 25%:

$$(40\%)(25\%) = (0.40)(0.25) = 0.10 = 10\%$$

The answer is (A).

If you don't already at least half-know these formulas, they're probably not worth worrying about. But if you can remember them without too much trouble, one of them might come in handy on test day.

There's a lot of material in this chapter, but remember that it's just the tip of the content pyramid. You can use the Miscellaneous Topics Follow-Up Test to see how much you've picked up from this chapter, but keep it in perspective. It's a higher priority to master the material in the chapters preceding this one.

> ### PROBABILITY FORMULAS
>
> Probability =
> $$\frac{\text{\# of favorable outcomes}}{\text{Total \# of possible outcomes}}$$
>
> If the probability that an event will occur is a, then the probability that the event will not occur is $1 - a$.
>
> If the probability that one event will occur is a and the independent probability that another event will occur is b, then the probability that both events will occur is ab.

THINGS TO REMEMBER:

Rules of Imaginary Numbers

$ai + bi = (a + b)i$
$ai - bi = (a - b)i$
$(ai)(bi) = -ab$
$i^2 = -1$
$i^3 = -i$
$i^4 = 1$

Complex Numbers

A number in the form $a + bi$, where a and b are real numbers, is called a complex number.

Contrapositive

"If p, then q," is logically equivalent to "If not q, then not p."

Average, Median, and Mode

Average (arithmetic mean) $= \dfrac{\text{Sum of the terms}}{\text{Number of terms}}$

Median = middle value (or average of two middle values)

Mode = most frequent value

Sum of the terms = (Average) × (Number of terms)

Percent Increase and Decrease

Percent increase $= \dfrac{\text{Amount of increase}}{\text{Original Amount}} \times 100\%$

Percent decrease $= \dfrac{\text{Amount of decrease}}{\text{Original Amount}} \times 100\%$

Average Rate Formula

Average A per $B = \dfrac{\text{Total } A}{\text{Total } B}$

Average speed $= \dfrac{\text{Total Distance}}{\text{Total Time}}$

Permutations and Combinations

The number of permutations of n distinct objects is

$n! = n(n - 1)(n - 2)\cdots(3)(2)(1)$

The number of permutations of n objects, a of which are indistinguishable, b of which are indistinguishable, etc., is

$\dfrac{n!}{a!b!\ldots}$

The number of permutations of n objects taken r at a time is

$nP_r = \dfrac{n!}{(n-r)!}$

The number of combinations of n objects taken r at a time is

$nC_r = \dfrac{n!}{r!(n-r)!}$

Probability

Probability $= \dfrac{\text{\# of favorable outcomes}}{\text{Total \# of possible outcomes}}$

If the probability that an event will occur is a, then the probability that the event will not occur is $1 - a$.

If the probability that one event will occur is a and the independent probability that another event will occur is b, then the probability that both events will occur is ab.

MISCELLANEOUS TOPICS FOLLOW-UP TEST

9 Questions (11 Minutes)

Directions: Solve problems 1–9. Fill in the oval corresponding to the best answer choice in the grid to the right of each question. (Answers and explanations begin on page 168.)

1. The "geocenter" of two positive numbers is defined as the positive square root of their product. If the geocenter of 5 and x is 9, what is the value of x ?

   (A) 6.7   (B) 7.0   (C) 11.3   (D) 13.0   (E) 16.2

   Ⓐ Ⓑ Ⓒ Ⓓ Ⓔ

2. "If A is true, then B is false," is logically equivalent to which of the following?

   I. If A is false, then B is true.
   II. If B is false, then A is true.
   III. If B is true, then A is false.

   (A)   I only
   (B)   III only
   (C)   I and II only
   (D)   II and III only
   (E)   I, II, and III   Ⓐ Ⓑ Ⓒ Ⓓ Ⓔ

3. If $i^2 = -1$, then all of the following are equal EXCEPT

   (A) $-i^2$ (B) $(-i)^2$ (C) i^4 (D) $(-i)^4$ (E) $-i^6$

   Ⓐ Ⓑ Ⓒ Ⓓ Ⓔ

KAPLAN

4. To convert a temperature reading from degrees Fahrenheit to degrees Regis, multiply the Fahrenheit reading by $\frac{5}{12}$ and subtract 216 from the result. Which of the following represents the Fahrenheit reading that is equivalent to a Regis reading of R degrees?

   (A) $\frac{5}{12}(R + 90)$

   (B) $\frac{5}{12}(R - 216)$

   (C) $\frac{5}{12}(R + 216)$

   (D) $\frac{12}{5}(R - 216)$

   (E) $\frac{12}{5}(R + 216)$ Ⓐ Ⓑ Ⓒ Ⓓ Ⓔ

5. The original price P of a certain item is first discounted by 20 percent, and then 5 percent of the discount price is added for sales tax. If the final price, including the sales tax, is $71.40, what was the original price P?

   (A) $59.50
   (B) $81.40
   (C) $84.00
   (D) $85.00
   (E) $86.40   Ⓐ Ⓑ Ⓒ Ⓓ Ⓔ

6. The average (arithmetic mean) of all the grades on a certain algebra test was 90. If the average of the 8 males' grades was 87, and the average of the females' grades was 92, how many females took the test?

   (A) 8   (B) 9   (C) 10   (D) 11   (E) 12

                  Ⓐ Ⓑ Ⓒ Ⓓ Ⓔ

DO YOUR FIGURING HERE

7. Rachel drives one-third of the distance from A to B at x kilometers per hour, and she drives the other two-thirds of the distance at y kilometers per hour. What is Rachel's average rate of speed, in kilometers per hour and in terms of x and y, for the entire trip?

(A) $\dfrac{x+y}{2}$

(B) $\dfrac{x+2y}{3}$

(C) $\dfrac{2x+y}{3}$

(D) $\dfrac{xy}{x+y}$

(E) $\dfrac{3xy}{2x+y}$ Ⓐ Ⓑ Ⓒ Ⓓ Ⓔ

8. How many ways are there to assign three people to five desks, with no more than one person to a desk?

(A) 8   (B) 15   (C) 30   (D) 60   (E) 120

Ⓐ Ⓑ Ⓒ Ⓓ Ⓔ

9. A bag contains 10 balls, each labeled with a different integer from 1 to 10, inclusive. If 2 balls are drawn simultaneously from the bag at random, what is the probability that the sum of the integers on the balls drawn will be greater than 6?

(A) 0.41   (B) 0.43   (C) 0.60   (D) 0.76   (E) 0.87

Ⓐ Ⓑ Ⓒ Ⓓ Ⓔ

DO YOUR FIGURING HERE

KAPLAN

FOLLOW-UP TEST—ANSWERS AND EXPLANATIONS

1. E

If 9 is the geocenter of 5 and x, then 9 is equal to the positive square root of the product of 5 and x:

$$9 = \sqrt{5x}$$
$$9^2 = \left(\sqrt{5x}\right)^2$$
$$81 = 5x$$
$$x = \frac{81}{5} = 16.2$$

2. B

If we think about this logic question in the general form "If p, then q," as described on pages 156-157, statement I is not necessarily true: It just negates p and q. (The negation of "B is false," is "B is true.") Statement II is not necessarily true: It just switches p and q. Statement III is true: It's the contrapositive.

3. B

Evaluate each choice:

(A) $-i^2 = (-1)(i^2) = (-1)(-1) = 1$
(B) $(-i)^2 = (-i)(-i) = i^2 = -1$
(C) $i^4 = (i^2)(i^2) = (-1)(-1) = 1$
(D) $(-i)^4 = (-i)^2(-i)^2 = (i^2)(i^2) = (-1)(-1) = 1$
(E) $-i^6 = (-1)(i^6) = (-1)(i^2)(i^2)(i^2) = (-1)(-1)(-1)(-1) = 1$

4. E

Translate carefully:

$$R = \frac{5}{12}F - 216$$

Then solve for F in terms of R:

$$R + 216 = \frac{5}{12}F$$
$$F = \frac{12}{5}(R + 216)$$

5. D

After a 20% discount, the original price P goes down to $0.8P$. That price is then increased by 5%, so the final price is $1.05(0.8P)$. The final price is given as \$71.40, so you can set up an equation and solve for the original price P:

$$1.05(0.8P) = 71.40$$
$$0.84P = 71.40$$
$$P = \frac{71.40}{0.84} = 85.00$$

6. E

The class average is not simply the average of the males' average and the females' average. The class average will be "weighted" in the direction of the larger subgroup. The class average (90) is closer to the female average (92) than to the male average (87), so there must be more females than males. But how many exactly?

As usual with averages questions, the key is to use the sum. The sum of the 8 males' scores is

$$\text{Male sum} = (\text{Average})(\text{Number})$$
$$= 87 \times 8$$
$$= 696$$

The sum of the x females' scores is

$$\text{Female sum} = (\text{Average})(\text{Number})$$
$$= 92x$$

The sum of all the scores is the class average (90) times the total number of people (8 males and x females)

$$\text{Class sum} = (\text{Average})(\text{Number})$$
$$= 90(8 + x)$$
$$= 720 + 90x$$

Now set the sum of the male sum and the female sum equal to the class sum and solve for x:

$$(\text{Male sum}) + (\text{Female sum}) = \text{Class sum}$$
$$696 + 92x = 720 + 90x$$
$$2x = 24$$
$$x = 12$$

7. E

To get the average speed, you need the total distance and the total time. Whatever you call the distance, it will drop out eventually, so you can call it whatever you want—a letter, an expression, a number. You might as well make it something easy to work with. The question stem speaks of "one-third" and "two-thirds" the distance, so it would be easiest to pick a total distance like $3d$ kilometers. (Another good total distance you might have is 300 kilometers.) For the first d kilometers, Rachel drives at x kilometers per hour.

Distance equals rate times time, so time equals distance over rate, and the time for the first leg is $\dfrac{d}{x}$ hours.

For the next $2d$ kilometers, Rachel drives at y kilometers per hour, so the time for the second leg is $\dfrac{2d}{y}$ hours.

$$
\begin{aligned}
\text{Average rate} &= \frac{\text{Total distance}}{\text{Total time}}\\[6pt]
&= \frac{3d}{\dfrac{d}{x}+\dfrac{2d}{y}}\\[6pt]
&= \frac{3d}{\dfrac{dy}{xy}+\dfrac{2dx}{xy}}\\[6pt]
&= \frac{3d}{\dfrac{dy+2dx}{xy}}\\[6pt]
&= \frac{3d}{1}\times\frac{xy}{dy+2dx}\\[6pt]
&= \frac{3dxy}{dy+2dx}\\[6pt]
&= \frac{3xy}{2x+y}
\end{aligned}
$$

8. D

This is a permutations question with a twist. Think of it as the arrangement of five items: three distinct people and two indistinguishable nonpersons.

$$
\begin{aligned}
\text{Number of permutations} &= \frac{n!}{n-r!}\\[6pt]
&= \frac{5!}{2!}\\[6pt]
&= \frac{5\times4\times3\times2\times1}{2\times1}\\[6pt]
&= 5\times4\times3\\[6pt]
&= 60
\end{aligned}
$$

9. E

To find the probability, you need to know the number of favorable outcomes and the total number of possible outcomes. The latter is the number of combinations of 10 items taken 2 at a time:

$$
{}_{10}C_2 = \frac{10!}{8!\,2!} = \frac{10\times9}{2} = 45
$$

To find the number of favorable outcomes, it's easiest to find the number of unfavorable outcomes (that is, where the numbers add up to 6 or less) by listing and counting them. These are the only unfavorable combinations:

1 and 2	1 and 5
1 and 3	2 and 3
1 and 4	2 and 4

All the other $45 - 6 = 39$ outcomes are favorable, and the probability is $39 \div 45 = 0.87$.

Part Three

PRACTICE TESTS

HOW TO TAKE THE PRACTICE TESTS

Before taking a practice test, find a quiet room where you can work uninterrupted for one hour. Make sure you have several No. 2 pencils with erasers.

Use the answer grid provided to record your answers. Guidelines for scoring your test appear on the reverse side of the answer grid. Time yourself. Spend no more than one hour on the 50 questions. Once you start the practice test, don't stop until you've reached the one-hour time limit. You'll find an answer key and complete answer explanations following the test. Be sure to read the explanations for all questions, even those you answered correctly.

Good luck!

HOW TO CALCULATE YOUR SCORE

Step 1: Figure out your raw score. Use the answer key to count the number of questions you answered correctly and the number of questions you answered incorrectly. (Do not count any questions you left blank.) Multiply the number wrong by 0.25 and subtract the result from the number correct. Round the result to the nearest whole number. This is your raw score.

SAT Subject Test: Mathematics Level 1 — Practice Test 1

Number right		Number wrong		Raw score

$$\boxed{} - \left(0.25 \times \boxed{}\right) = \boxed{}$$

Step 2: Find your scaled score. In the Score Conversion Table below, find your raw score (rounded to the nearest whole number) in one of the columns to the left. The score directly to the right of that number will be your scaled score.

A note on your practice test scores: Don't take these scores too literally. Practice test conditions cannot precisely mirror real test conditions. Your actual SAT Subject Test: Mathematics Level 1 score will almost certainly vary from your practice test scores. However, your scores on the practice tests will give you a rough idea of your range on the actual exam.

Conversion Table

Raw	Scaled	Raw	Scaled	Raw	Scaled	Raw	Scaled
50	800	34	610	18	460	2	320
49	790	33	600	17	450	1	310
48	780	32	590	16	450	0	310
47	770	31	580	15	440	−1	300
46	750	30	570	14	430	−2	290
45	740	29	560	13	420	−3	280
44	730	28	550	12	400	−4	270
43	720	27	540	11	390	−5	260
42	700	26	530	10	390	−6	260
41	690	25	520	9	380	−7	250
40	680	24	510	8	370	−8	240
39	670	23	500	7	360	−9	230
38	660	22	500	6	350	−10	230
37	650	21	490	5	350	−11	220
36	640	20	480	4	340	−12	210
35	620	19	470	3	330		

Answer Grid
Practice Test 1

1. Ⓐ Ⓑ Ⓒ Ⓓ Ⓔ
2. Ⓐ Ⓑ Ⓒ Ⓓ Ⓔ
3. Ⓐ Ⓑ Ⓒ Ⓓ Ⓔ
4. Ⓐ Ⓑ Ⓒ Ⓓ Ⓔ
5. Ⓐ Ⓑ Ⓒ Ⓓ Ⓔ
6. Ⓐ Ⓑ Ⓒ Ⓓ Ⓔ
7. Ⓐ Ⓑ Ⓒ Ⓓ Ⓔ
8. Ⓐ Ⓑ Ⓒ Ⓓ Ⓔ
9. Ⓐ Ⓑ Ⓒ Ⓓ Ⓔ
10. Ⓐ Ⓑ Ⓒ Ⓓ Ⓔ
11. Ⓐ Ⓑ Ⓒ Ⓓ Ⓔ
12. Ⓐ Ⓑ Ⓒ Ⓓ Ⓔ
13. Ⓐ Ⓑ Ⓒ Ⓓ Ⓔ
14. Ⓐ Ⓑ Ⓒ Ⓓ Ⓔ
15. Ⓐ Ⓑ Ⓒ Ⓓ Ⓔ
16. Ⓐ Ⓑ Ⓒ Ⓓ Ⓔ
17. Ⓐ Ⓑ Ⓒ Ⓓ Ⓔ
18. Ⓐ Ⓑ Ⓒ Ⓓ Ⓔ
19. Ⓐ Ⓑ Ⓒ Ⓓ Ⓔ
20. Ⓐ Ⓑ Ⓒ Ⓓ Ⓔ
21. Ⓐ Ⓑ Ⓒ Ⓓ Ⓔ
22. Ⓐ Ⓑ Ⓒ Ⓓ Ⓔ
23. Ⓐ Ⓑ Ⓒ Ⓓ Ⓔ
24. Ⓐ Ⓑ Ⓒ Ⓓ Ⓔ
25. Ⓐ Ⓑ Ⓒ Ⓓ Ⓔ

26. Ⓐ Ⓑ Ⓒ Ⓓ Ⓔ
27. Ⓐ Ⓑ Ⓒ Ⓓ Ⓔ
28. Ⓐ Ⓑ Ⓒ Ⓓ Ⓔ
29. Ⓐ Ⓑ Ⓒ Ⓓ Ⓔ
30. Ⓐ Ⓑ Ⓒ Ⓓ Ⓔ
31. Ⓐ Ⓑ Ⓒ Ⓓ Ⓔ
32. Ⓐ Ⓑ Ⓒ Ⓓ Ⓔ
33. Ⓐ Ⓑ Ⓒ Ⓓ Ⓔ
34. Ⓐ Ⓑ Ⓒ Ⓓ Ⓔ
35. Ⓐ Ⓑ Ⓒ Ⓓ Ⓔ
36. Ⓐ Ⓑ Ⓒ Ⓓ Ⓔ
37. Ⓐ Ⓑ Ⓒ Ⓓ Ⓔ
38. Ⓐ Ⓑ Ⓒ Ⓓ Ⓔ
39. Ⓐ Ⓑ Ⓒ Ⓓ Ⓔ
40. Ⓐ Ⓑ Ⓒ Ⓓ Ⓔ
41. Ⓐ Ⓑ Ⓒ Ⓓ Ⓔ
42. Ⓐ Ⓑ Ⓒ Ⓓ Ⓔ
43. Ⓐ Ⓑ Ⓒ Ⓓ Ⓔ
44. Ⓐ Ⓑ Ⓒ Ⓓ Ⓔ
45. Ⓐ Ⓑ Ⓒ Ⓓ Ⓔ
46. Ⓐ Ⓑ Ⓒ Ⓓ Ⓔ
47. Ⓐ Ⓑ Ⓒ Ⓓ Ⓔ
48. Ⓐ Ⓑ Ⓒ Ⓓ Ⓔ
49. Ⓐ Ⓑ Ⓒ Ⓓ Ⓔ
50. Ⓐ Ⓑ Ⓒ Ⓓ Ⓔ

right

wrong

Use the answer key following the test to count up the number of questions you got right and the number you got wrong. (Remember not to count omitted questions as wrong.) "How to Calculate Your Score" on the previous page will show you how to find your score.

Practice Test 1

50 Questions (1 hour)

Directions: For each question, choose the BEST answer from the choices given. If the precise answer is not among the choices, choose the one that best approximates the answer. Then fill in the corresponding oval on the answer sheet.

Notes:

(1) To answer some of these questions, you will need a calculator. You must use at least a scientific calculator, but programmable and graphing calculators are also allowed.

(2) All angle measures on this test are in degrees, so your calculator should be set to degree mode.

(3) Figures in this test are drawn as accurately as possible UNLESS it is stated in a specific question that the figure is not drawn to scale. All figures are assumed to lie in a plane unless otherwise specified.

(4) The domain of any function f is assumed to be the set of all real numbers x for which $f(x)$ is a real number, unless otherwise indicated.

Reference Information: Use the following formulas as needed.

Right circular cone: If r = radius and h = height, then Volume = $\frac{1}{3}\pi r^2 h$, and if c = circumference of the base and ℓ = slant height, then Lateral Area = $\frac{1}{2}c\ell$.

Sphere: If r = radius, then Volume = $\frac{4}{3}\pi r^3$ and Surface Area = $4\pi r^2$.

Pyramid: If B = area of the base and h = height, then Volume = $\frac{1}{3}Bh$.

1.   If $2a + 3 = 6$, then $\dfrac{3}{4a + 6} =$

(A) $\dfrac{1}{4}$ (B) $\dfrac{1}{2}$ (C) 1   (D) 2   (E) 3

2.   In terms of x, what is the average (arithmetic mean) of $4x - 2$, $x + 2$, $2x + 3$, and $x + 1$?

(A) $2x - 1$
(B) $2x$
(C) $2x + 1$
(D) $2x + 4$
(E) $8x + 4$

3.   If $4^{2x+2} = 64$, then $x =$

(A) $\dfrac{1}{2}$ (B) 1   (C) $\dfrac{3}{2}$ (D) 2   (E) $\dfrac{5}{2}$

4.   What is the least positive integer that is divisible by both 2 and 5 and leaves a remainder of 2 when it is divided by 7?

(A) 20   (B) 30   (C) 50   (D) 65   (E) 75

5.   In Figure 1, the area of rectangle $CDEF$ is twice the area of rectangle $ABCF$. If $CD = 2x + 2$, what is the length of $\overline{AE}$, in terms of x?

(A) $2x + 3$
(B) $2x + 4$
(C) $3x + 1$
(D) $3x + 2$
(E) $3x + 3$

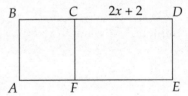

Figure 1

GO ON TO THE NEXT PAGE

6. If $2y^2 + x - 4 = 0$ and $\dfrac{x}{2} = y^2$, then $x =$

(A) 1   (B) 2   (C) 3   (D) 4   (E) 5

7. If a laser printer can print x pages per minute, how many minutes, in terms of x, would it take the laser printer to print a 100-page document?

(A) $100x$

(B) $100 - x$

(C) $100 + x$

(D) $\dfrac{x}{100}$

(E) $\dfrac{100}{x}$

8. In the table, $f(x)$ is a linear function. What is the value of k?

(A)   3
(B)   4
(C)   5
(D)   6
(E)   7

x	$f(x)$
0	–4
1	–1
2	2
3	k
4	8

9. Jackie uses 30 percent of her monthly earnings for rent and 50 percent of the remaining amount for food and transportation. If she spends \$525 for food and transportation, how much does she pay for rent?

(A) \$400   (B) \$450   (C) \$500   (D) \$550   (E) \$600

GO ON TO THE NEXT PAGE

KAPLAN

10. In Figure 2, if congruent right triangles ABD and DCA share leg $\overline{AD}$, then $x =$

(A)   90
(B)   100
(C)   110
(D)   120
(E)   130

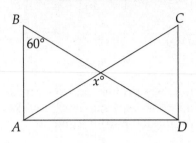

Figure 2

11. If $\dfrac{x+1}{2} + \dfrac{4x-1}{4} = 5.5$, then $x =$

(A) 2.5   (B) 3.0   (C) 3.5   (D) 4.0   (E) 4.5

12. If $a \downarrow b = \sqrt[b]{a}$, then $10 \downarrow 3 \approx$

(A) 1.12   (B) 1.69   (C) 2.15   (D) 2.71   (E) 3.33

13. Which of the following ordered pairs is the solution to the equations $2y + x = 5$ and $-2y + x = 9$?

(A) $(-7,-1)$ (B) $(-1, 7)$ (C) $(7, -1)$ (D) $(-7, 1)$ (E) $(1, 7)$

14. What is the solution set for the equation $|2x - 3| = 13$?

(A) $\{-8\}$
(B) $\{-5\}$
(C) $\{-5, -8\}$
(D) $\{-5, 8\}$
(E) $\{5, -8\}$

15. $\dfrac{6!}{2!3!} =$

(A) 1   (B) 6   (C) 15   (D) 30   (E) 60

GO ON TO THE NEXT PAGE

16. In Figure 3, the length of $\overline{AC}$ is 3 times the length of $\overline{CD}$. If B is the midpoint of $\overline{AC}$, and the length of $\overline{CD}$ is 5, what is the length of $\overline{BD}$?

    (A)   10
    (B)   12.5
    (C)   13.5
    (D)   15
    (E)   17.5

DO YOUR FIGURING HERE.

Figure 3

17. Which of the following lines is parallel to $y = -2x + 3$ and has a y-intercept of 4 ?

    (A) $y = -2x + 4$

    (B) $y = -2x - 4$

    (C) $y = 2x - 4$

    (D) $y = 2x + 4$

    (E) $y = \dfrac{1}{2}x + 4$

18. In Figure 4, the area of quadrilateral $ABCD$ is

    (A)   32
    (B)   33
    (C)   34
    (D)   35
    (E)   36

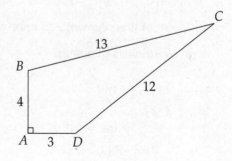

Figure 4

19. If $f(x) = x^2 + x$ and $g(x) = \sqrt{x}$, then $f(g(3)) \approx$

    (A)   1.73
    (B)   3.46
    (C)   4.73
    (D)   7.34
    (E)   12.00

GO ON TO THE NEXT PAGE

KAPLAN

20. At a certain software company, the cost, C, of developing and producing a computer software program is related to the number of copies produced, x, by the equation $C = 30{,}000 + 2x$. The company's total revenues, R, are related to the number of copies produced, x, by the equation $R = 6x - 10{,}000$. How many copies must the company produce so that the total revenue is equal to the cost?

(A)   5,000
(B)   6,000
(C)   7,500
(D)   9,000
(E)   10,000

21. If the two squares shown in Figure 5 are identical, what is the degree measure of angle ADE ?

(A)   120
(B)   135
(C)   150
(D)   165
(E)   175

22. Points $A\left(\sqrt{2},4\right)$, $B\left(6,-\sqrt{3}\right)$, and C are collinear. If B is the midpoint of line segment AC, approximately what are the (x, y) coordinates of point C ?

(A)   (3.71, 1.13)
(B)   (3.71, 5.73)
(C)   (7.41, –7.46)
(D)   (10.59, –7.46)
(E)   (10.59, 5.73)

DO YOUR FIGURING HERE.

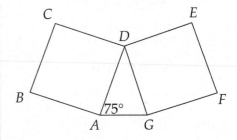

Figure 5

GO ON TO THE NEXT PAGE

KAPLAN

23.   What is the solution set to the equation $4 + x^2 = 2x^2 - 5$?

(A) $\{x: x = 3\}$
(B) $\{x: x = -3\}$
(C) $\{x: x = \pm 3\}$
(D) $\{x: x = -1\}$
(E) $\{x: x = 1\}$

DO YOUR FIGURING HERE.

24.   Which of the following triplets can be the lengths of the sides of a triangle?

(A)   2, 3, 5
(B)   1, 4, 2
(C)   7, 4, 4
(D)   5, 6, 12
(E)   9, 20, 8

25.   In Figure 6, if $\sin x = 0.500$, what is the approximate value of $\tan x$?

(A)   0.577
(B)   0.707
(C)   1.000
(D)   1.155
(E)   2.000

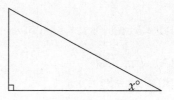

Figure 6

26.   In Figure 7, if line ℓ has a slope of 1 and passes through the origin, which of the following points has (x, y) coordinates such that $\dfrac{x}{y} > 1$?

(A) A
(B) B
(C) C
(D) D
(E) E

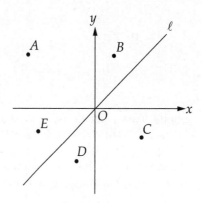

Figure 7

GO ON TO THE NEXT PAGE

KAPLAN

27. On a recent chemistry test, the average (arithmetic mean) score among 5 students was 83, where the lowest and highest possible scores were 0 and 100, respectively. If the teacher decides to increase each student's score by 2 points, and if none of the students originally scored more than 98, which of the following must be true?

    I. After the scores are increased, the average score is 85.

    II. When the scores are increased, the difference between the highest and lowest scores increases.

    III. After the increase, all 5 scores are greater than or equal to 25.

(A) I only
(B) II only
(C) I and II only
(D) I and III only
(E) I, II, and III

DO YOUR FIGURING HERE.

28. If $a > b$ and $c > d$, which of the following must be true?

(A) $ac > bd$
(B) $a + b > c + d$
(C) $a + c > b + d$
(D) $a - b > c - d$
(E) $ad > bc$

29. $1 - 2\sin^2\theta - 2\cos^2\theta =$

(A) –2   (B) –1   (C) 0   (D) 1   (E) 2

30. Sheila leaves her house and starts driving due south for 30 miles, then drives due west for 60 miles, and finally drives due north for 10 miles to reach her office. Which of the following is the approximate straight-line distance, in miles, from her house to her office?

(A) 63   (B) 67   (C) 71   (D) 75   (E) 80

GO ON TO THE NEXT PAGE

31. If $f(x) = x^2 - 1$, $g(x) = (x - 1)^{-1}$, and $x \neq 1$, then $f(x)g(x) =$

(A) $2x + 1$
(B) $x + 1$
(C) $x - 1$
(D) $x^3 - 1$
(E) $2x - 1$

DO YOUR FIGURING HERE.

32. If an empty rectangular water tank that has dimensions 100 centimeters, 20 centimeters, and 40 centimeters is to be filled using a right cylindrical bucket with a base radius of 9 centimeters and a height of 20 centimeters, approximately how many buckets of water will it take to fill the tank?

(A) 14   (B) 16   (C) 18   (D) 20   (E) 22

33. Sarah is scheduling the first four periods of her school day. She needs to fill those periods with calculus, art, literature, and physics, and each of these courses is offered during each of the first four periods. How many different schedules can Sarah choose from?

(A) 1   (B) 4   (C) 12   (D) 24   (E) 120

34. In Figure 8, $\overline{AE}$ is parallel to $\overline{BD}$. What is the approximate length of $\overline{DE}$?

(A) 2.33   (B) 2.67   (C) 3.33   (D) 3.67   (E) 6.67

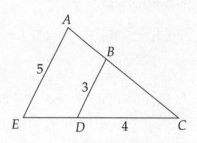

Figure 8

35. If $f(x) = \sqrt{x^2 - 4}$, what is the domain?

(A)   All real numbers
(B)   All x such that $x \geq 2$
(C)   All x such that $x \leq -2$
(D)   All x such that $-2 \leq x \leq 2$
(E)   All x such that $x \leq -2$ or $x \geq 2$

GO ON TO THE NEXT PAGE

KAPLAN

36. What is the area of a triangle with vertices (1,1), (3,1), and (5,7)?

    (A) 6   (B) 7   (C) 9   (D) 10   (E) 12

DO YOUR FIGURING HERE.

37. Which of the following inequalities is equivalent to $-2(x + 5) < -4$?

    (A) $x > -3$
    (B) $x < -3$
    (C) $x > 3$
    (D) $x < 3$
    (E) $x > 7$

38. If $i = \sqrt{-1}$, for which of the following values of n does $i^n + (-i)^n$ have a positive value?

    (A) 23   (B) 24   (C) 25   (D) 26   (E) 27

39. The maximum value of the function $f(x) = 1 - \cos x$ between 0 and 2π is

    (A) 0   (B) 1   (C) 1.5   (D) 2   (E) 2.5

40. In Figure 9, if $y > 60$ and $AB = BC$, which of the following must be true?

    I. $a + y = 180$

    II. $y > z$

    III. $a = y + z$

    (A)   I only
    (B)   II only
    (C)   III only
    (D)   I and II
    (E)   II and III

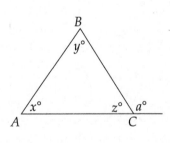

Figure 9

GO ON TO THE NEXT PAGE

41.   If $f(x) = \dfrac{1}{x}$, and $0 < x < 1$, what is the range of $f(x)$?

   (A)   All real numbers
   (B)   All real numbers between 0 and 1
   (C)   All real numbers greater than 0
   (D)   All real numbers greater than 1
   (E)   All real numbers greater than or equal to 1

42.   Two identical spheres of radius 6 intersect so that the distance between their centers is 10. The points of intersection of the two spheres form a circle. What is the area of this circle?

   (A) 5π (B) 6π (C) 8π (D) 10π (E) 11π

43.   If a remainder of 4 is obtained when $x^3 + 2x^2 - x - k$ is divided by $x - 2$, what is the value of k?

   (A) 4   (B) 6   (C) 10   (D) 12   (E) 14

44.   Ms. Hobbes has a portfolio that includes $50,000 in stock, $75,000 in cash, and no other holdings. If she wishes to redistribute her holdings so that 80 percent of the portfolio is in cash, how many dollars of stock must she convert to cash?

   (A)   10,000
   (B)   15,000
   (C)   20,000
   (D)   25,000
   (E)   30,000

DO YOUR FIGURING HERE.

GO ON TO THE NEXT PAGE

KAPLAN

45. A sphere of radius 5 has the same volume as a cube with an edge of approximately what length?

   (A)   5.00
   (B)   5.50
   (C)   6.24
   (D)   8.06
   (E)   9.27

46. The equation $x^2 = y^2$ is represented by which of the following graphs?

   (A)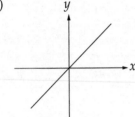

   (B)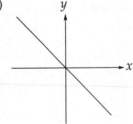

   (C)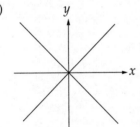

   (D)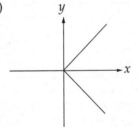

   (E)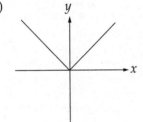

DO YOUR FIGURING HERE.

GO ON TO THE NEXT PAGE

KAPLAN

47. If point $A(3,5)$ is located on a circle in the coordinate plane, and the center of the circle is the origin, which of the following points must lie outside this circle?

    (A) $(1.0, 6.0)$
    (B) $(1.5, 5.5)$
    (C) $(2.5, 4.5)$
    (D) $(4.0, 4.0)$
    (E) $(5.0, 3.0)$

48. If $f(x) = 3x - 1$, n represents the slope of the line with the equation $y = f^{-1}(x)$, and p represents the slope of a line that is perpendicular to the line with the equation $y = f(x)$, then $np =$

    (A) -9

    (B) $-\dfrac{1}{9}$

    (C) $\dfrac{1}{9}$

    (D) 9

    (E)   It cannot be determined from the information given.

49. The parabola with the equation $y = 4x - \dfrac{1}{2}x^2$ has how many points with (x, y) coordinates that are both positive integers?

    (A)   3
    (B)   4
    (C)   7
    (D)   8
    (E)   Infinitely many

DO YOUR FIGURING HERE.

GO ON TO THE NEXT PAGE

50. In Figure 10, each of the three circles is tangent to the other two, and each side of the equilateral triangle is tangent to two of the circles. If the length of one side of the triangle is x, what is the radius, in terms of x, of one of the circles?

(A) $\dfrac{x}{1+2\sqrt{3}}$

(B) $\dfrac{x}{2+2\sqrt{3}}$

(C) $\dfrac{x}{1+\sqrt{3}}$

(D) $\dfrac{2x}{1+\sqrt{3}}$

(E) $\dfrac{2x}{1+2\sqrt{3}}$

DO YOUR FIGURING HERE.

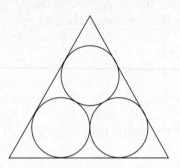

Figure 10

STOP!

If you finish before time is up, you may check your work.

**Turn the page
for answers and explanations
to Practice Test 1.**

Answer Key

1. A	18. E	35. E
2. C	19. C	36. A
3. A	20. E	37. A
4. B	21. A	38. B
5. E	22. D	39. D
6. B	23. C	40. E
7. E	24. C	41. D
8. C	25. A	42. E
9. B	26. E	43. C
10. D	27. D	44. D
11. C	28. C	45. D
12. C	29. B	46. C
13. C	30. A	47. A
14. D	31. B	48. B
15. E	32. B	49. A
16. B	33. D	50. B
17. A	34. B	

ANSWERS AND EXPLANATIONS

1. A

You don't need to find the value of a to answer this question. It's more direct to recognize that the denominator of the expression you're solving for is exactly twice the left side of the given equation:

$$2a + 3 = 6$$
$$2(2a + 3) = 2(6)$$
$$4a + 6 = 12$$
$$\frac{1}{4a + 6} = \frac{1}{12}$$
$$\frac{3}{4a + 6} = \frac{3}{12} = \frac{1}{4}$$

2. C

To find the average of four expressions, add them up and divide by 4. The sum is $8x + 4$. When you divide that by 4, you get $2x + 1$.

3. A

To solve an equation with the unknown in an exponent, rewrite the equation so that both sides have the same base:

$$4^{2x+2} = 64$$
$$4^{2x+2} = 4^3$$
$$2x + 2 = 3$$
$$2x = 1$$
$$x = \frac{1}{2}$$

4. B

The easiest way to find the answer to this question is to check the answer choices, starting with the smallest, until you find one that meets the given conditions. A multiple of 2 and 5 is a multiple of 10, so already you know that the answer is (A), (B), or (C). Starting with (A), divide each by 7 until you find one that leaves a remainder of 2. (A) 20 divided by 7 is 2 with a remainder of 6—that's not it. (B) 30 divided by 7 is 4 with a remainder of 2—that's it.

5. E

The two rectangles have the same height, so if $CDEF$ has twice the area of $ABCF$, then the base of $CDEF$ is twice the base of $ABCF$. The longer base is given as $2x + 2$, so the shorter base is half that, or $x + 1$. You're looking for AE, which is the sum of the lengths of the bases:

$$AE = AF + FE$$
$$= (x + 1) + (2x + 2)$$
$$= 3x + 3$$

6. B

The second equation expresses y^2 in terms of x, so you can substitute that expression for y^2 in the first equation and solve for x :

$$2y^2 + x - 4 = 0$$
$$2\left(\frac{x}{2}\right) + x - 4 = 0$$
$$x + x - 4 = 0$$
$$2x = 4$$
$$x = 2$$

7. E

A rate of x pages per minute is the same as 100 pages per how many minutes? If y is the number of minutes it takes to print 100 pages, then you can set up the proportion:

$$\frac{x \text{ pages}}{1 \text{ minute}} = \frac{100 \text{ pages}}{y \text{ minutes}}$$
$$xy = 100$$
$$y = \frac{100}{x}$$

8. C

The given values of x are evenly spaced, and $f(x)$ is a linear (i.e., straight-line) function, so the values of $f(x)$ will be evenly spaced. The unknown k is halfway between 2 and 8, so $k = 5$.

KAPLAN

9. B

After spending 30% on rent, Jackie has 70% of her earnings left. Half of that 70% is 35%, which goes for food and transportation. So the ratio of rent to food and transportation is 30:35, or 6:7. Now you can set up a proportion and solve for the rent:

$$\frac{\text{rent}}{\text{food \& trans}} = \frac{6}{7}$$

$$\frac{\text{rent}}{\$525} = \frac{6}{7}$$

$$\text{rent} = \frac{6}{7} \times \$525 = 6 \times \$75$$

$$= \$450$$

10. D

Mark up the figure. Because the triangles are congruent, $\angle C$ also measures 60°. And because they're both right triangles, $\angle BDA$ and $\angle CAD$ both measure 30°:

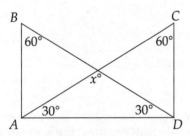

Now you have the other two angles inside a triangle with x, so you can set their sum equal to 180 and solve for x:

$$x + 30 + 30 = 180$$

$$x = 120$$

11. C

To solve an equation with fractions, first multiply both sides by whatever you have to to clear all denominators. In this case, multiply both sides by 4:

$$\frac{x+1}{2} + \frac{4x-1}{4} = 5.5$$

$$4\left(\frac{x+1}{2} + \frac{4x-1}{4}\right) = 4(5.5)$$

$$(2x+2) + (4x-1) = 22$$

$$6x + 1 = 22$$

$$6x = 21$$

$$x = \frac{21}{6} = 3.5$$

12. C

Plug $a = 10$ and $b = 3$ into the definition:

$$a \downarrow b = \sqrt[b]{a}$$

$$10 \downarrow 3 = \sqrt[3]{10} \approx 2.15$$

13. C

If you add the equations as presented, the y's drop out:

$$2y + x = 5$$
$$\underline{-2y + x = 9}$$
$$2x = 14$$
$$x = 7$$

The only answer choice that works is (C). To confirm, plug $x = 7$ back into one of the original equations to find y:

$$2y + x = 5$$
$$2y + 7 = 5$$
$$2y = -2$$
$$y = -1$$

14. D

To solve an equation with absolute value, consider the two possibilities.

If $|2x - 3| = 13$, then what's inside the absolute value signs equals either 13 or –13:

$$2x - 3 = 13$$
$$2x = 16$$
$$x = 8$$

OR:

$$2x - 3 = -13$$
$$x = -10$$
$$x = -5$$

15. E

Expand the factorials, cancel factors common to the numerator and denominator, and then calculate:

$$\frac{6!}{2!3!} = \frac{6 \times 5 \times 4 \times 3 \times 2 \times 1}{2 \times 1 \times 3 \times 2 \times 1}$$
$$= \frac{6 \times 5 \times 4}{2} = 60$$

16. B

Mark up the figure. $CD = 5$ and AC is 3 times that, so $\overline{AC} = 15$:

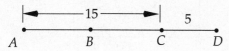

B is the midpoint of $\overline{AC}$, so $AB = BC = 7.5$:

```
      7.5        7.5        5
   •----------•----------•------•
   A          B          C      D
```

Now you can see that $BD = BC + CD = 7.5 + 5 = 12.5$.

17. A

Parallel lines have the same slope, so you're looking for a line with a slope of –2. Only (A) and (B) have that slope, and of those, only (A) has y-intercept +4.

18. E

Add to the figure. Diagonal BD will divide the quadrilateral into two familiar triangles:

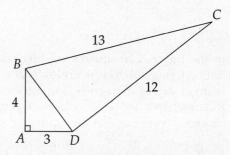

ABD is a right triangle with legs 3 and 4, so $BD = 5$, and therefore BCD is a 5-12-13 right triangle.

The area of the 3-4-5 triangle is

$$\frac{1}{2}(\text{leg}_1)(\text{leg}_2) = \frac{1}{2}(3)(4) = 6$$

And the area of the 5-12-13 right triangle is

$$\frac{1}{2}(\text{leg}_1)(\text{leg}_2) = \frac{1}{2}(5)(12) = 30$$

The quadrilateral's area, then, is $6 + 30 = 36$

19. C

Apply the inside function first:

$$g(x) = \sqrt{x}$$
$$g(3) = \sqrt{3}$$

Then apply the outside function to the result:

$$f(x) = x^2 + x$$
$$f(\sqrt{3}) = (\sqrt{3})^2 + \sqrt{3}$$
$$= 3 + \sqrt{3} \approx 4.73$$

20. E

Put much more simply, what the question is asking is: At what point does cost equal revenue? In algebraic terms, for what value of x does C equal R ?

$$C = R$$
$$30,000 + 2x = 6x - 10,000$$
$$-4x = -40,000$$
$$x = 10,000$$

21. A

Mark up the figure. The squares are identical, so $AD = DG$ and triangle ADG is isosceles. That makes the measure of $\angle DGA$ 75 degrees, and the measure of $\angle ADG$ is therefore $180° - 75° - 75° = 30$ degrees. Now you know every angle in the figure:

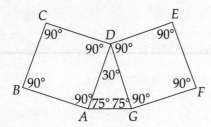

The measure of $\angle ADE$ is $30° + 90° = 120$ degrees.

22. D

If $B(6, -\sqrt{3})$ is the midpoint of $A(\sqrt{2}, 4)$ and $C(x, y)$, then 6 is the average of $\sqrt{2}$ and x, and $-\sqrt{3}$ is the average of 4 and y :

$$\frac{\sqrt{2} + x}{2} = 6$$
$$\sqrt{2} + x = 12$$
$$x = 12 - \sqrt{2} \approx 10.59$$

and:

$$\frac{4 + y}{2} = -\sqrt{3}$$
$$4 + y = -2\sqrt{3}$$
$$y = -2\sqrt{3} - 4 \approx -7.46$$

23. C

Simplify the equation:

$$4 + x^2 = 2x^2 - 5$$
$$4 + 5 = 2x^2 - x^2$$
$$9 = x^2$$

Don't jump to the conclusion that if 9 equals x^2, then x must be 3. x could just as well be –3:

$$x^2 = 9$$
$$x = \pm 3$$

24. C

To satisfy the Triangle Inequality Theorem, adding the two shorter sides must give you more than the longest side. In (B), (D), and (E), the two shorter sides add up to less than the longest side, and in (A), they add up to the same as the longest side. Only in (C) do they add up to more.

25. A

If sin x is 0.5, then the side opposite x is one-half the hypotenuse. In other words, this is a 30-60-90 triangle with familiar side ratios:

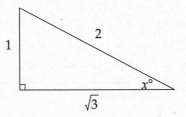

Tangent is opposite over adjacent, so

$$\tan x = \frac{1}{\sqrt{3}} \approx 0.577$$

26. E

When x and y have different signs, (that is, when the point is in the second or fourth quadrant), $\frac{x}{y}$ is negative, so for no point in those quadrants can $\frac{x}{y}$ be greater than 1. That eliminates (A) and (C). When x and y are both positive, $\frac{x}{y}$ will be greater than 1 if x is greater than y. Point B is above the line $x = y$, so for that point $y > x$. When x and y are both negative, $\frac{x}{y}$ will be greater than 1 if x is less than y. Point D is below the line $x = y$, so for that point $y < x$. Point E, however, is above the line $x = y$, so for that point $y > x$, and $\frac{x}{y} > 1$.

27. D

If the average of the 5 original scores was 83, then the sum of those 5 scores was $5 \times 83 = 415$. When each score is increased by 2, the sum goes up to 425, one-fifth of which is 85. So the new average is 85 and statement I is true. Statement II, however, is not true because if both the lowest and highest scores go up by the same amount, the difference between those scores remains the same. So I is true and II is not, and the answer is either (A) or (D). What about III? To find the lowest score that one of the 5 students can get, imagine that after the increase the other 4 students all get perfect 100s. Four 100s add up to 400, which leaves $425 - 400 = 25$ for the fifth and lowest possible score for the 5 students. III is true.

28. C

You could do this one by picking numbers, but for any set of numbers you might pick, more than one of the answer choices will be true. You will therefore have to pick at least two sets of numbers to find the correct answer. The statement that must be true is the one that holds for all a, b, c, and d that fit the given conditions. (A) is tempting: You might think that the product of the larger a and c will be greater than the product of the smaller b and d, and it is if all you pick are positive integers. (A) doesn't hold when you consider negatives, however. But (C) holds for all possible values of $a, b, c,$ and d. The sum of two larger quantities is greater than the sum of two smaller quantities.

29. B

Look for "$(\sin^2 + \cos^2)$" in the expression:

$$1 - 2\sin^2\theta - 2\cos^2\theta = 1 - 2(\sin^2\theta + \cos^2\theta)$$

$$= 1 - 2 = -1$$

30. A

Sketch a diagram:

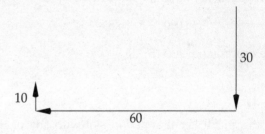

The net change is south 20 and west 60:

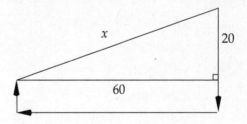

So you're looking for the hypotenuse of a right triangle with legs of 20 and 60:

$$x = \sqrt{20^2 + 60^2}$$

$$= \sqrt{400 + 3,600} = \sqrt{4,000} \approx 63$$

31. B

This time there's no inside and outside functions. What you're looking for here is the product of the functions:

$$f(x) = x^2 - 1$$

$$g(x) = (x - 1)^{-1} = \frac{1}{x - 1}$$

$$f(x)g(x) = (x^2 - 1)\left(\frac{1}{x - 1}\right)$$

$$= \frac{x^2 - 1}{x - 1}$$

$$= \frac{(x - 1)(x + 1)}{x - 1}$$

$$= x + 1$$

32. B

The volume of the tank is 100 cm $\times$ 20 cm $\times$ 40 cm = 80,000 cubic centimeters. The volume of the bucket is $\pi r^2 h \approx (3.14)(9^2)(20) = 5,086.8$. To figure out how many buckets fit into the tank, divide:

$$80,000 \div 5,086.8 \approx 15.7$$

The nearest choice is (B), 16.

33. D

Sarah can fill her first-period slot with any of the 4 subjects, leaving 3 for the second, 2 for the third, and 1 for the fourth:

$$4 \times 3 \times 2 \times 1 = 24$$

34. B

That $\overline{AE}$ and $\overline{BD}$ are parallel tells you that the outside triangle ACE and the inside triangle BCD are similar. Corresponding sides are proportional:

$$\frac{BD}{AE} = \frac{CD}{CE}$$

$$\frac{3}{5} = \frac{4}{4 + x}$$

$$3(4 + x) = 5 \times 4$$

$$12 + 3x = 20$$

$$3x = 8$$

$$x = \frac{8}{3} \approx 2.67$$

35. E

For what values of x can you take the square root of $x^2 - 4$? Only those values for which $x^2 - 4$ is nonnegative:

$$x^2 - 4 \geq 0$$

$$x^2 \geq 4$$

$$x \leq -2 \text{ or } x \geq 2$$

36. A

Sketch the diagram:

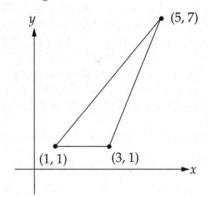

If you use the short side from (1, 1) to (3, 1) as the base, then the base is 2 and the height is 6:

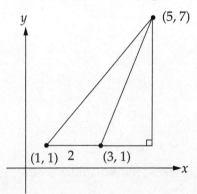

Plug $b = 2$ and $h = 6$ into the formula for the area of a triangle:

$$\text{Area of Triangle} = \frac{1}{2}bh = \frac{1}{2}(2)(6) = 6$$

37. A

Simplify:

$$-2(x + 5) < -4$$

$$-2x - 10 < -4$$

$$-2x < -4 + 10$$

$$-2x < 6$$

$$x > \frac{6}{-2}$$

$$x > -3$$

Notice that the inequality sign flipped when both sides were divided by –2.

38. B

When you raise i to successive integer exponents, a pattern develops:

$$i^1 = i$$

$$i^2 = -1$$

$$i^3 = -i$$

$$i^4 = 1$$

$$i^5 = i$$

Every fourth power the pattern repeats. A similar pattern develops with $-i$:

$$(-i)^1 = -i$$

$$(-i)^2 = -1$$

$$(-i)^3 = i$$

$$(-i)^4 = 1$$

$$(-i)^5 = -i$$

Whenever the exponent is a multiple of 4, i^n and $(-i)^n$ are both equal to 1. (B) is the only multiple of 4 among the choices. When $n = 24$, $i^n + (-i)^n = 1 + 1 = 2$, which is a positive integer.

39. D

Between 0 and 2π, cos x ranges from –1 to 1. The maximum value of $f(x) = 1 - \cos x$ will be when cos x is at its least; that is, when cos $x = -1$:

$$f(x) = 1 - \cos x$$

$$= 1 - (-1) = 1 + 1 = 2$$

40. E

The triangle is isosceles; since $AB = BC$, $x = z$. Since $y > 60$, that leaves less than 120 degrees to split between x and z, so x and z are both less than 60. Angle a is an exterior angle, so it's equal to the sum of the remote interior angles:

$$a = x + y$$

Now you can see that II is true, because $y > 60$ and $z < 60$, and you can see that III is true, because $a = x + y$ and $x = z$. I is not true: y is greater than z and $a + z = 180$, so $a + y$ is greater than 180.

41. D

Imagine (or use your graphing calculator) the graph of $f(x) = \dfrac{1}{x}$ between 0 and 1. At $x = 1$, $f(x) = 1$. As x gets smaller, $f(x)$ gets bigger: $f\left(\dfrac{1}{2}\right) = 2$, $f\left(\dfrac{1}{100}\right) = 100$. As x approaches 0, $f(x)$ gets infinitely large. The range is all real numbers greater than 1.

42. E

This one's hard to sketch because it's 3-D. Here's a cross section:

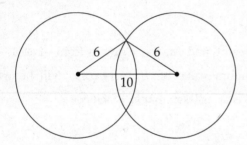

Drop an altitude, and you make two right triangles:

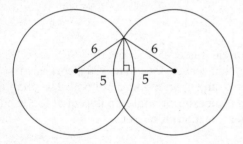

That altitude is the radius of the circle formed by the cross section of the spheres. (Can you picture it?)

$$r = \sqrt{6^2 - 5^2}$$
$$= \sqrt{36 - 25}$$
$$= \sqrt{11}$$
$$\pi r^2 = 11\pi$$

43. C

$$
\begin{array}{r}
x^2 + 4x + 7 \\
x - 2 \overline{)\, x^3 + 2x^2 - x - k} \\
\underline{x^3 - 2x^2} \\
4x^2 - x \\
\underline{4x^2 - 8x} \\
7x - k \\
\underline{7x - 14} \\
-k + 14
\end{array}
$$

In algebraic terms, the remainder is $-k + 14$. If that's equal to 4, then

$$-k + 14 = 4$$
$$10 = k$$

44. D

The total portfolio is $50{,}000 + 75{,}000 = 125{,}000$ dollars. Eighty percent of that is $(0.80)(125{,}000) = 100{,}000$ dollars. Thus, she needs to convert $100{,}000 - 75{,}000 = 25{,}000$ dollars.

45. D

A sphere of radius 5 has volume $\dfrac{4}{3}\pi(5)^3 \approx 523.6$. If that's also the volume of a cube, then an edge of that cube is the cube root of that. Use your calculator: $\sqrt[3]{523.6} \approx 8.06$.

46. C

The equation $x^2 = y^2$ is true whenever $x = \pm y$. Choice (A) shows $x = y$. Choice (B) shows $x = -y$. Choice (C) shows both.

47. A

$\overline{OA}$ is a radius of the circle, and

$$OA = \sqrt{3^2 + 5^2} = \sqrt{34}$$

You're looking for the choice for which $x^2 + y^2$ is greater than 34. Use your calculator. In (A) $x^2 + y^2 = 37$. So it looks like the answer is (A). If you have lots of time, you can check the other answer choices, just to be sure. In (B) $x^2 + y^2 = 32.5$. In (C) $x^2 + y^2 = 26.5$. In (D) $x^2 + y^2 = 32$. And in (E) $x^2 + y^2 = 34$. Only (A) is greater than 34.

48. B

You don't actually need to find the inverse function to answer this question if you remember that the slopes of inverse functions are reciprocals. The slope of the given $f(x) = 3x - 1$ is 3, so the slope of $f^{-1}(x)$ is $\frac{1}{3}$. The slope of a line perpendicular to $f(x)$ will have a slope that's the negative reciprocal of 3, or $-\frac{1}{3}$. So $n = \frac{1}{3}$ and $p = -\frac{1}{3}$, and therefore $np = \frac{1}{3}\left(-\frac{1}{3}\right) = -\frac{1}{9}$.

49. A

The question is asking, in other words, for how many positive integer values of x does y turn out to be a positive integer?

The expression $4x - \frac{1}{2}x^2$ will be positive whenever $4x$ is greater than $\frac{1}{2}x^2$:

$$4x > \frac{1}{2}x^2$$

$$\frac{1}{2}x^2 - 4x < 0$$

$$x\left(\frac{1}{2}x - 4\right) < 0$$

$$x > 0 \text{ and } \frac{1}{2}x - 4 < 0$$

$$x > 0 \text{ and } x < 8$$

$$0 < x < 8$$

The expression $4x - \frac{1}{2}x^2$ will be an integer whenever $\frac{1}{2}x^2$ is an integer.

Half of x^2 is an integer whenever x is even, so y will be a positive integer if x is an even number greater than 0 and less than 8—in other words, if x is 2, 4, or 6. That's 3 points.

50. B

Mark up the figure:

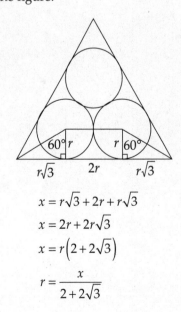

$$x = r\sqrt{3} + 2r + r\sqrt{3}$$
$$x = 2r + 2r\sqrt{3}$$
$$x = r\left(2 + 2\sqrt{3}\right)$$
$$r = \frac{x}{2 + 2\sqrt{3}}$$

HOW TO CALCULATE YOUR SCORE

Step 1: Figure out your raw score. Use the answer key to count the number of questions you answered correctly and the number of questions you answered incorrectly. (Do not count any questions you left blank.) Multiply the number wrong by 0.25 and subtract the result from the number correct. Round the result to the nearest whole number. This is your raw score.

SAT Subject Test: Mathematics Level 1 — Practice Test 2

Number right	Number wrong	Raw score

$$\boxed{} - \left(0.25 \times \boxed{}\right) = \boxed{}$$

Step 2: Find your scaled score. In the Score Conversion Table below, find your raw score (rounded to the nearest whole number) in one of the columns to the left. The score directly to the right of that number will be your scaled score.

A note on your practice test scores: Don't take these scores too literally. Practice test conditions cannot precisely mirror real test conditions. Your actual SAT Subject Test: Mathematics Level 1 score will almost certainly vary from your practice test scores. However, your scores on the practice tests will give you a rough idea of your range on the actual exam.

Conversion Table

Raw	Scaled	Raw	Scaled	Raw	Scaled	Raw	Scaled
50	800	34	610	18	460	2	320
49	790	33	600	17	450	1	310
48	780	32	590	16	450	0	310
47	770	31	580	15	440	−1	300
46	750	30	570	14	430	−2	290
45	740	29	560	13	420	−3	280
44	730	28	550	12	400	−4	270
43	720	27	540	11	390	−5	260
42	700	26	530	10	390	−6	260
41	690	25	520	9	380	−7	250
40	680	24	510	8	370	−8	240
39	670	23	500	7	360	−9	230
38	660	22	500	6	350	−10	230
37	650	21	490	5	350	−11	220
36	640	20	480	4	340	−12	210
35	620	19	470	3	330		

Answer Grid
Practice Test 2

1. (A) (B) (C) (D) (E)   26. (A) (B) (C) (D) (E)

2. (A) (B) (C) (D) (E)   27. (A) (B) (C) (D) (E)

3. (A) (B) (C) (D) (E)   28. (A) (B) (C) (D) (E)

4. (A) (B) (C) (D) (E)   29. (A) (B) (C) (D) (E)

5. (A) (B) (C) (D) (E)   30. (A) (B) (C) (D) (E)

6. (A) (B) (C) (D) (E)   31. (A) (B) (C) (D) (E)

7. (A) (B) (C) (D) (E)   32. (A) (B) (C) (D) (E)

8. (A) (B) (C) (D) (E)   33. (A) (B) (C) (D) (E)

9. (A) (B) (C) (D) (E)   34. (A) (B) (C) (D) (E)

10. (A) (B) (C) (D) (E)   35. (A) (B) (C) (D) (E)

11. (A) (B) (C) (D) (E)   36. (A) (B) (C) (D) (E)

12. (A) (B) (C) (D) (E)   37. (A) (B) (C) (D) (E)

13. (A) (B) (C) (D) (E)   38. (A) (B) (C) (D) (E)

14. (A) (B) (C) (D) (E)   39. (A) (B) (C) (D) (E)

15. (A) (B) (C) (D) (E)   40. (A) (B) (C) (D) (E)

16. (A) (B) (C) (D) (E)   41. (A) (B) (C) (D) (E)

17. (A) (B) (C) (D) (E)   42. (A) (B) (C) (D) (E)

18. (A) (B) (C) (D) (E)   43. (A) (B) (C) (D) (E)

19. (A) (B) (C) (D) (E)   44. (A) (B) (C) (D) (E)

20. (A) (B) (C) (D) (E)   45. (A) (B) (C) (D) (E)

21. (A) (B) (C) (D) (E)   46. (A) (B) (C) (D) (E)

22. (A) (B) (C) (D) (E)   47. (A) (B) (C) (D) (E)

23. (A) (B) (C) (D) (E)   48. (A) (B) (C) (D) (E)

24. (A) (B) (C) (D) (E)   49. (A) (B) (C) (D) (E)

25. (A) (B) (C) (D) (E)   50. (A) (B) (C) (D) (E)

right

wrong

Use the answer key following the test to count up the number of questions you got right and the number you got wrong. (Remember not to count omitted questions as wrong.) "How to Calculate Your Score" on the previous page will show you how to find your score.

Practice Test 2

50 Questions (1 hour)

Directions: For each question, choose the BEST answer from the choices given. If the precise answer is not among the choices, choose the one that best approximates the answer. Then fill in the corresponding oval on the answer sheet.

Notes:

(1) To answer some of these questions, you will need a calculator. You must use at least a scientific calculator, but programmable and graphing calculators are also allowed.

(2) All angle measures on this test are in degrees, so your calculator should be set to degree mode.

(3) Figures in this test are drawn as accurately as possible UNLESS it is stated in a specific question that the figure is not drawn to scale. All figures are assumed to lie in a plane unless otherwise specified.

(4) The domain of any function f is assumed to be the set of all real numbers x for which $f(x)$ is a real number, unless otherwise indicated.

Reference Information: Use the following formulas as needed.

Right circular cone: If r = radius and h = height, then Volume = $\frac{1}{3}\pi r^2 h$; and if c = circumference of the base and ℓ = slant height, then Lateral Area = $\frac{1}{2}c\ell$.

Sphere: If r = radius, then Volume = $\frac{4}{3}\pi r^3$ and Surface Area = $4\pi r^2$.

Pyramid: If B = area of the base and h = height, then Volume = $\frac{1}{3}Bh$.

---

1. If $a + b = 10$ and both a and b are positive integers, what is the largest possible value for a ?

    (A) 6   (B) 7   (C) 8   (D) 9   (E) 10

DO YOUR FIGURING HERE.

2. If $2x + y = 2y - x$, then $y =$

    (A) $3x$
    (B) $4x$
    (C) $2x + 2$
    (D) $3x + 1$
    (E) $4x + 1$

3. If a certain car can travel 240 miles on 12 gallons of gasoline, then at the same rate, how many gallons of gasoline are needed to travel 300 miles?

    (A) 10   (B) 15   (C) 20   (D) 25   (E) 30

4. If $2(1 + 2x) - 5(4 - 2x) = 14$, $x \approx$

    (A) –5.33   (B) –0.29   (C) 0.29   (D) 0.67   (E) 2.29

5. In Figure 1, if $\overline{ABC}$ is a straight line, $x =$

    (A)   50
    (B)   60
    (C)   70
    (D)   80
    (E)   90

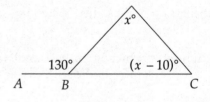

Figure 1

6. $\left|-3.1\right| - \left|-4.2\right| + \left|2.3\right| =$

    (A) 1.2   (B) 3.4   (C) 6.9   (D) 7.3   (E) 9.6

GO ON TO THE NEXT PAGE

KAPLAN)

7.   If $x \neq 0$, then $\dfrac{x+1}{6x} + \dfrac{x+1}{2x} =$

     (A) $\dfrac{2x+2}{3x}$

     (B) $\dfrac{2x+2}{4x}$

     (C) $\dfrac{2x+3}{6x}$

     (D) $\dfrac{2x+2}{8x}$

     (E) $\dfrac{x+2}{6x}$

DO YOUR FIGURING HERE.

8.   If $8 = \dfrac{1}{\dfrac{1}{x^{-3}}}$, then $x =$

     (A) -2.00
     (B) -0.50
     (C) 0.50
     (D) 1.25
     (E) 2.00

9.   Which of the following is the equation of a line that makes a 45-degree angle with the x-axis and has a y-intercept of 2?

     (A) $y = x + 2$
     (B) $y = x - 2$
     (C) $y = 45x + 2$
     (D) $y = 45x - 2$
     (E) $y = 2x + 45$

GO ON TO THE NEXT PAGE

KAPLAN

10.   If x and y are both positive, x is even, and y is odd, which of the following must be odd?

DO YOUR FIGURING HERE.

(A) xy

(B) $x + 2y$

(C) x^y

(D) y^x

(E) $\dfrac{x}{y}$

11.   In Figure 2, if the length of $\overline{AB}$ is 2 more than 3 times the length of $\overline{BC}$, and $AC = 14$, what is the length of $\overline{BC}$?

A $\qquad\qquad\qquad\qquad$ B $\quad$ C

Figure 2

(A)   2
(B)   3
(C)   4
(D)   7
(E)   12

12.   Jean can paint a house in 10 hours, and Dan can paint the same house in 12 hours. If Jean begins the job and does $\dfrac{1}{3}$ of it and then Dan takes over and finishes the job, what is the total time it takes them to paint the house?

(A)   10 hours, 40 minutes
(B)   11 hours
(C)   11 hours, 20 minutes
(D)   11 hours, 40 minutes
(E)   12 hours

GO ON TO THE NEXT PAGE

13. If $a = 4b$, $c = 8b^2$, and $b \neq 0$, then $\dfrac{c - a}{4b} =$

(A) $-2b$

(B) $1 - 2b$

(C) $2b$

(D) $2b - 1$

(E) $2b + 3$

14. In Figure 3, if O is the center of the circle, and the ratio of x to y is 2 to 1, what is the ratio of a to b?

(A) 4:1

(B) 3:1

(C) 2:1

(D) 1:2

(E) 1:4

15. The cost of 2 sodas and 3 pretzels is $4.60. If the cost of 2 pretzels is $2.20, what is the cost of 2 sodas?

(A) $0.60

(B) $0.65

(C) $1.20

(D) $1.30

(E) $2.40

DO YOUR FIGURING HERE.

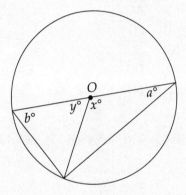

Figure 3

GO ON TO THE NEXT PAGE

KAPLAN

16. If $\dfrac{2}{x}\left(x^2 + x\right) = \dfrac{1}{2}$, then $x + 1 =$

(A) $\dfrac{1}{4}$ (B) $\dfrac{1}{2}$ (C) 1   (D) $\dfrac{3}{2}$ (E) 2

DO YOUR FIGURING HERE.

17. In Figure 4, if 5 lines are drawn by connecting the labeled points with the origin, which of the lines would have the greatest slope?

(A) $\overline{AO}$
(B) $\overline{BO}$
(C) $\overline{CO}$
(D) $\overline{DO}$
(E) $\overline{EO}$

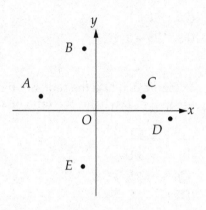

Figure 4

18. If $f(x) = 5x - x^2$, then $f(x) = 6$ when $x =$

(A)   1 only
(B)   2 only
(C)   1 or 5
(D)   2 or 3
(E)   2 or 6

19. If 25 percent of a certain number is 36, what would 40 percent of the same number be?

(A)   13.6
(B)   22.5
(C)   42.6
(D)   52.8
(E)   57.6

GO ON TO THE NEXT PAGE

20. In Figure 5, if the area of square *ABCD* is 5, what is the area of square *BEFD* ?

    (A)   7.07
    (B)   8.25
    (C)   10.00
    (D)   12.50
    (E)   25.00

21. If $\sqrt{x} + 6 = x$, then $x =$

    (A)   4 only
    (B)   9 only
    (C)   4 or 9
    (D)   −4 or 9
    (E)   There is no solution.

22. A rubber ball is dropped from a height of 10 meters. If the ball always rebounds $\frac{4}{5}$ the distance it has fallen, how far, in meters, will the ball have traveled at the moment it hits the ground for the fourth time?

    (A)   4.10
    (B)   5.12
    (C)   29.52
    (D)   43.92
    (E)   49.04

DO YOUR FIGURING HERE.

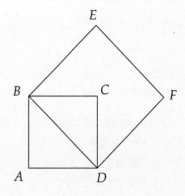

Figure 5

GO ON TO THE NEXT PAGE

23. If a varies directly as b and if $a = 4$ when $b = 5$, what is the value of a when $b = 10$?

    (A) 4   (B) 5   (C) 6   (D) 8   (E) 9

24. Which of the following figures has the greatest number of lines of symmetry?

    (A)   Equilateral triangle
    (B)   Rectangle
    (C)   Square
    (D)   Rhombus
    (E)   Circle

25. If $f(x) = x^2 - 4x + 1$, $f(x)$ crosses the x-axis closest to which of the following points?

    (A) $(-0.33, 0)$
    (B) $(0.27, 0)$
    (C) $(1.73, 0)$
    (D) $(3.27, 0)$
    (E) $(4.33, 0)$

DO YOUR FIGURING HERE.

GO ON TO THE NEXT PAGE

KAPLAN

26. If the graph in Figure 6 represents $f(x)$, which of the follow-ing graphs would represent $|f(x)|$?

DO YOUR FIGURING HERE.

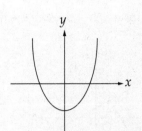

Figure 6

(A)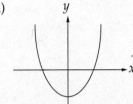

(B)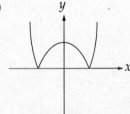

(C)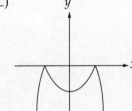

(D)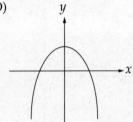

(E)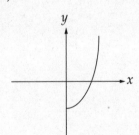

27. If $(x + y)^2 = (x - y)^2$, which of the following must be true?

(A) $x = y$

(B) $x = -y$

(C) $x = 0$

(D) $y = 0$

(E) $x = 0$ or $y = 0$

GO ON TO THE NEXT PAGE

KAPLAN

28.   Which of the following diagrams represents the solution set for $y \geq 2x + 3$ and $y \leq -\dfrac{1}{2}x + 1$?

DO YOUR FIGURING HERE.

(A)

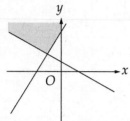

(B)

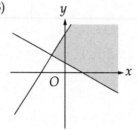

(C)

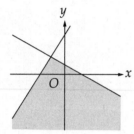

(D)

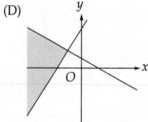

(E)

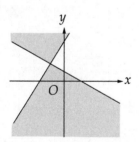

29.   If $x \lozenge y = \sqrt{x^2 - y^2}$, then $10 \lozenge (-6) =$

(A) 2   (B) 8   (C) 16   (D) 32   (E) 64

GO ON TO THE NEXT PAGE

KAPLAN

30. In Figure 7, if *ABCDEF* is a regular hexagon, what is the approximate length of $\overline{AE}$ if *AB* = 4 ?

   (A)   5.65
   (B)   5.73
   (C)   6.00
   (D)   6.93
   (E)   8.00

DO YOUR FIGURING HERE.

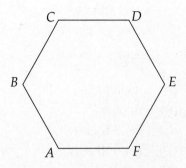

Figure 7

31. Which of the following is an equation of a line that will never intersect with the line that has the equation $y = 4x + 2$?

   (A) $y + 4x = 3$

   (B) $2y + 4x = 2$

   (C) $y - 4x = -3$

   (D) $y + 4x = -2$

   (E) $y + \dfrac{1}{4} = 3$

32. If $i = \sqrt{-1}$, then $(3 + i)(3 - i) =$

   (A) 8   (B) 9   (C) 10   (D) $9 - i$ (E) $9 + i$

GO ON TO THE NEXT PAGE

33.   In Figure 8, which of the following must be true?

I. $\sin x = \dfrac{3}{5}$

II. $\tan y = \tan r$

III. $\cos x = \sin s$

(A)   I only
(B)   III only
(C)   I and II only
(D)   I and III only
(E)   I, II, and III

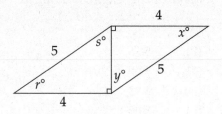

Figure 8

34.   Figure 9 shows a semicircle that is the graph of the equation $y = \sqrt{6x - x^2}$. If the semicircle is rotated 360° about the x–axis, what is the volume of the sphere that is created?

(A) 6π
(B) 12π
(C) 18π
(D) 24π
(E) 36π

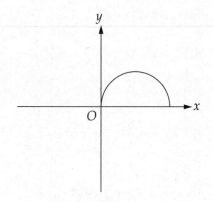

Figure 9

35.   If $f(x) = 2x$ and $f(f(x)) = x + 1$, then $x =$

(A) $\dfrac{1}{3}$ (B) 1   (C) 2   (D) 3   (E) 5

GO ON TO THE NEXT PAGE

36. In Figure 10, if $\ell_1 \parallel \ell_2$, $x =$

    (A)   15
    (B)   25
    (C)   30
    (D)   35
    (E)   40

37. Sally is interested in buying a car. If she has a choice of 3 colors (red, green, or blue), 2 body types (two or four doors), and 3 engine types (four, six, or eight cylinders), how many different models can she choose from?

    (A) 6   (B) 8   (C) 12   (D) 16   (E) 18

38. Which of the following is the solution set for $|\,2x\,| < 4$ and $|\,x - 2\,| < 2$?

    (A) $-2 < x < 0$
    (B) $-2 < x < 2$
    (C) $0 < x < 2$
    (D) $0 < x < 4$
    (E) $2 < x < 4$

39. If when $f(x)$ is divided by $3x + 1$, the quotient is $x^2 - x + 3$ and the remainder is 2, then $f(x) =$

    (A) $3x^3 - 2x^2 - 8x + 3$
    (B) $3x^3 - 2x^2 + 8x + 5$
    (C) $3x^3 + 4x^2 + 8x + 1$
    (D) $3x^3 - 4x^2 + 8x - 1$
    (E) $3x^3 - 4x^2 + 8x - 3$

DO YOUR FIGURING HERE.

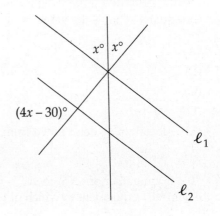

Figure 10

GO ON TO THE NEXT PAGE

40.   Figure 11 shows an equilateral triangle with two of its vertices on sides of a square and its third vertex on a vertex of the square. What is the value of $y - x$?

(A)   45

(B)   60

(C)   75

(D)   90

(E)   It cannot be determined from the information given.

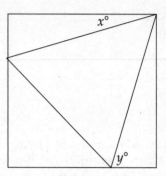

Figure 11

41.   "If x is a member of set S, then x is not a member of set T," is logically equivalent to which of the following?

   I. If x is a member of set T, then x is not a member of set S.

   II. If x is not a member of set T, then x is a member of set S.

   III. If x is not a member of set S, then x is a member of set T.

(A)   I only

(B)   II only

(C)   III only

(D)   I and II

(E)   II and III

42.   In Figure 12, if triangle ABC is an isosceles triangle of perimeter 20, what is the approximate area of the circle with center O ?

(A)   26.83

(B)   33.65

(C)   44.17

(D)   57.33

(E)   229.34

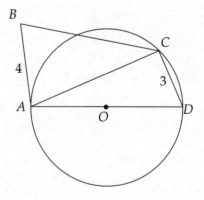

Figure 12

GO ON TO THE NEXT PAGE

KAPLAN

43. If P is a point on the line $y = 2x$ in the first quadrant, and the distance from the origin to point P is 5, what are the approximate coordinates of point P ?

    (A)   (2.24, 4.47)
    (B)   (3.00, 6.00)
    (C)   (4.00, 8.00)
    (D)   (4.47, 2.24)
    (E)   (4.72, 2.36)

44. In Figure 13, circle O has diameter $\overline{AB}$ of length 8. If smaller circle P is tangent to diameter $\overline{AB}$ at point O and is also tangent to circle O, what is the approximate area of the shaded region?

    (A)   3.14
    (B)   6.28
    (C)   9.42
    (D)   12.57
    (E)   25.13

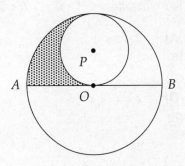

Figure 13

45. On a number line, the coordinate of point A is 0, and the coordinate of point B is 6. If point P is located on the number line so that the distance from P to A is twice the distance from P to B, which of the following could be the coordinate of point P ?

    (A)   2 only
    (B)   4 only
    (C)   2 or 4
    (D)   4 or 12
    (E)   12 only

46. How many different-sized rectangular solids with a volume of 32 cubic units are there such that each dimension has an integer value?

    (A) 4   (B) 5   (C) 6   (D) 10   (E) 12

GO ON TO THE NEXT PAGE

47. If the roots of the equation $2x^2 + k = 8x$ are real, which of the following expresses all the possible values for k ?

(A) $k \le 8$

(B) $k \ge 8$

(C) $k \ge 4$

(D) $k \le -8$

(E) $k \ge -8$

48. For what values of x is $x^2 + 6 < 5x$?

(A) $2 < x < 3$

(B) $-2 < x < 3$

(C) $-3 < x < -2$

(D) $-2 < x$ or $x > 3$

(E) $-3 < x$ or $x > 2$

49. In a cube with edge length 2, the distance from the center of one face to a vertex of the opposite face is

(A) 2.00

(B) 2.24

(C) 2.45

(D) 2.65

(E) 2.83

50. A certain stock begins the week trading at $87\frac{1}{2}$ per share. If the average gain for the next four days is $\frac{1}{2}$, by how much should the price of the stock increase during Friday so that the total gain for the stock during the entire five days is 5 percent?

(A) $1\frac{3}{4}$ (B) $1\frac{7}{8}$ (C) $2\frac{1}{8}$ (D) $2\frac{1}{4}$ (E) $2\frac{3}{8}$

DO YOUR FIGURING HERE.

STOP!

If you finish before time is up, you may check your work.

KAPLAN

**Turn the page
for answers and explanations
to Practice Test 2.**

Answer Key

1. D	18. D	35. A
2. A	19. E	36. D
3. B	20. C	37. E
4. E	21. B	38. C
5. C	22. E	39. B
6. A	23. D	40. B
7. A	24. E	41. A
8. C	25. B	42. D
9. A	26. B	43. A
10. D	27. E	44. B
11. B	28. D	45. D
12. C	29. B	46. B
13. D	30. D	47. A
14. D	31. C	48. A
15. D	32. C	49. C
16. A	33. D	50. E
17. E	34. E	

ANSWERS AND EXPLANATIONS

1. D

For a to be as large as possible, make b as small as possible. If $b = 1$, then $a = 9$.

2. A

Isolate y:

$$2x + y = 2y - x$$
$$y - 2y = -x - 2x$$
$$-y = -3x$$
$$y = 3x$$

3. B

Set up a proportion:

$$\frac{240 \text{ miles}}{12 \text{ gallons}} = \frac{300 \text{ miles}}{x \text{ gallons}}$$
$$240x = 12 \times 300$$
$$x = 15$$

4. E

Solve for x:

$$2(1 + 2x) - 5(4 - 2x) = 14$$
$$2 + 4x - 20 + 10x = 14$$
$$14x - 18 = 14$$
$$14x = 32$$
$$x = \frac{32}{14} \approx 2.29$$

5. C

The angle marked 130° is an exterior angle and is equal to the sum of the remote interior angles marked $x°$ and $(x - 10)°$:

$$x + (x - 10) = 130$$
$$2x - 10 = 130$$
$$2x = 140$$
$$x = 70$$

6. A

Find the absolute values first, and then calculate:

$$|-3.1| - |-4.2| + |2.3| = 3.1 - 4.2 + 2.3$$
$$= -1.1 + 2.3$$
$$= 1.2$$

7. A

To add fractions, you need a common denominator. Here that would be $6x$. Multiply the top and bottom of the second fraction by 3 and you can proceed:

$$\frac{x+1}{6x} + \frac{x+1}{2x} = \frac{x+1}{6x} + \frac{3x+3}{6x}$$
$$= \frac{x+1+3x+3}{6x}$$
$$= \frac{4x+4}{6x}$$
$$= \frac{2x+2}{3x}$$

8.   C

The right side of this equation is the reciprocal of the reciprocal of x^{-3}:

$$8 = \frac{1}{\frac{1}{x^{-3}}}$$

$$8 = x^{-3}$$

$$\frac{1}{x^3} = 8$$

$$x^3 = \frac{1}{8}$$

$$x = \frac{1}{2} = 0.50$$

9.   A

To make a 45-degree angle with the x-axis, a line must have a slope of 1 or –1. Of the given choices, only (A) and (B) have such a slope, and of those, only (A) has a y-intercept of 2.

10.   D

(A) is always even because the product of an even and an odd is even. (B) is always even because the sum of an even and twice an odd is even. (C) is always even because no matter what positive integer exponent you raise an even number to, the result will be even. (D), on the other hand, is always odd because no matter what positive integer exponent you raise an odd number to, the result will be odd. (E) can never be odd because if an even divided by an odd gave you an integer at all, that integer would have to be even.

11.   B

Mark up the figure:

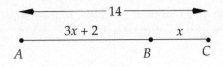

You can see that

$$(3x + 2) + x = 14$$

$$4x + 2 = 14$$

$$4x = 12$$

$$x = 3$$

12.   C

If Jean can do the whole job in 10 hours, then it takes her $\frac{1}{3} \times 10 = 3\frac{1}{3}$ hours to do $\frac{1}{3}$ of the job. If Dan can do the whole job in 12 hours, then it takes him $\frac{2}{3} \times 12 = 8$ hours to do the other $\frac{2}{3}$ of the job. Thus, together, it takes them $3\frac{1}{3} + 8 = 11\frac{1}{3}$ hours, or 11 hours and 20 minutes, to paint the house.

13.   D

Plug $a = 4b$ and $c = 8b^2$ into the expression:

$$\frac{c - a}{4b} = \frac{8b^2 - 4b}{4b}$$

$$= \frac{4b(2b - 1)}{4b}$$

$$= 2b - 1$$

14. D

Angles x and y are supplementary, so if $x + y = 180$ and $x = 2y$, then $x = 120$ and $y = 60$. Because all radii are equal, you can tell that the triangle with angles b and y is equilateral, so $b = 60$, and the triangle with angles a and x is isosceles with a vertex angle of $120°$, so $a = 30$. The ratio of a to b, then, is 1:2.

15. D

If the cost of 2 pretzels is \$2.20, then each pretzel costs \$1.10, and 3 pretzels cost \$3.30. Two sodas, then, cost \$4.60 − \$3.30 = \$1.30.

16. A

Notice that if you distribute just the $\frac{1}{x}$, you'll get an equation in terms of $x + 1$:

$$\frac{2}{x}\left(x^2 + x\right) = \frac{1}{2}$$

$$2\left(\frac{x^2 + x}{x}\right) = \frac{1}{2}$$

$$2(x + 1) = \frac{1}{2}$$

$$x + 1 = \frac{1}{4}$$

17. E

The more "uphill" the line (as you move from left to right), the greater the slope. Lines $\overrightarrow{AO}$, $\overrightarrow{BO}$, and $\overrightarrow{DO}$ will all be "downhill" lines—that is, they all have negative slopes. Lines $\overleftrightarrow{CO}$ and $\overleftrightarrow{EO}$ will be uphill—they have positive slopes—but $\overleftrightarrow{EO}$ is visibly "steeper" than $\overleftrightarrow{CO}$:

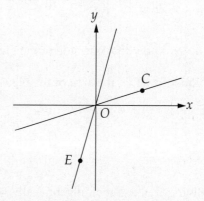

Line EO has the greatest slope.

18. D

If $f(x)$ equals both $5x - x^2$ and 6 for some particular value of x, then you can set $5x - x^2$ equal to 6 and solve for x:

$$5x - x^2 = 6$$
$$x^2 - 5x + 6 = 0$$
$$(x - 2)(x - 3) = 0$$
$$x = 2 \text{ or } 3$$

19. E

If 25% of a number is 36, then that number is 36 times 4, or 144. Forty percent of that is $(0.40)(144) = 57.6$.

20. C

The area of the smaller square is 5, so each side is $\sqrt{5}$. Diagonal $\overline{BD}$ divides the square into 45–45–90 triangles, so the hypotenuse BD equals one of the legs times $\sqrt{2}$:

$$BD = \left(\sqrt{5}\right)\left(\sqrt{2}\right) = \sqrt{10}$$

Now that you know that one side of the big square is $\sqrt{10}$, you know that the area of the big square is $\left(\sqrt{10}\right)^2 = 10$.

21. B

When solving this equation algebraically, you will square both sides of the equation, which can result in a false solution:

$$\sqrt{x} + 6 = x$$
$$\sqrt{x} = x - 6$$
$$\left(\sqrt{x}\right)^2 = (x - 6)^2$$
$$x = x^2 - 12x + 36$$
$$x^2 - 13x + 36 = 0$$
$$(x - 4)(x - 9) = 0$$
$$x = 4 \text{ or } 9$$

Try $x = 4$ and $x = 9$ in the original equation, and you'll find that $x = 4$ doesn't really work. The only solution is $x = 9$.

22. E

Sketch a diagram. First the ball drops 10 meters and rebounds $\dfrac{4}{5}$ of that, or 8 meters:

Then it drops 8 meters and rebounds $\dfrac{4}{5}$ of that, or 6.4 meters, and continuing this way you get this:

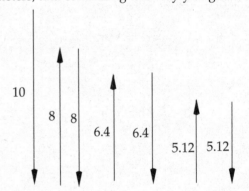

The sum of these 7 distances is

$$10 + 8 + 8 + 6.4 + 6.4 + 5.12 + 5.12 = 49.04$$

23. D

If *a* varies directly as *b*, then *a* is equal to some constant times *b*:

$$a = kb$$
$$4 = k(5)$$
$$k = 0.8$$

Now plug in *b* = 10 and *k* = 0.8:

$$a = kb = (0.8)(10) = 8$$

24. E

A line of symmetry is a line along which you can "fold" a figure such that every point on one side of the fold aligns with a point on the other side of the fold. An equilateral triangle has three lines of symmetry:

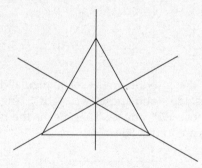

A nonsquare rectangle has two, a nonsquare rhombus has two, and a square has four:

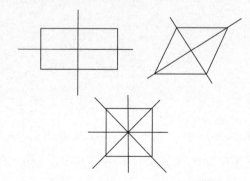

A circle, however, has infinitely many. Any fold through the center of the circle is a line of symmetry:

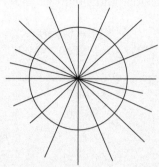

25. B

The graph crosses the *x*-axis at the point where *f*(*x*) = 0. So

$$x^2 - 4x + 1 = 0$$
$$x = \frac{4 \pm \sqrt{16 - 4}}{2}$$
$$= \frac{4 \pm \sqrt{12}}{2}$$
$$= 2 \pm \sqrt{3} \approx 3.73 \text{ or } 0.27$$

Choice (B) matches one of these.

26. B

The graph of $|f(x)|$ is the same as the graph of *f*(*x*) except that any part below the *x*-axis has to be flipped above the *x*-axis. That's what choice (B) shows.

27. E

Expand both sides and simplify:

$$(x+y)^2 = (x-y)^2$$
$$x^2 + 2xy + y^2 = x^2 - 2xy + y^2$$
$$2xy = -2xy$$
$$4xy = 0$$
$$xy = 0$$

Either $x = 0$ or $y = 0$.

28. D

The right sides of both inequalities are in $mx + b$ form, so it's easy to find the slopes and y-intercepts.

In the first inequality, the slope is 2 and the y-intercept is 3.

In the second inequality the slope is $-\frac{1}{2}$ (the negative reciprocal of the other slope, which makes the lines perpendicular) and the y-intercept is 1.

All choices have the lines in the right places. The difference is in the shading. The first inequality has a "greater-than-or-equal-to" sign, so to satisfy that inequality, a point must be on or above the line $y = 2x + 3$. So we can eliminate B, C, and E.

The second inequality has a "less-than-or-equal-to" sign, so to satisfy that inequality, a point must be on or below the line $y = -\frac{1}{2}x + 1$. The choice with the proper shading, then, is (D).

29. B

Plug $x = 10$ and $y = -6$ into the definition:

$$x \lozenge y = \sqrt{x^2 - y^2}$$
$$10 \lozenge (-6) = \sqrt{10^2 - (-6)^2} = \sqrt{100 - 36}$$
$$= \sqrt{64} = 8$$

30. D

Mark up the figure. Draw in not only the segment AE you're looking for, but also the perpendicular that makes the two right triangles as shown:

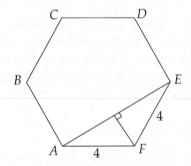

Each of the interior angles of a regular hexagon measures 120°, and these two right triangles split one of those 120° angles. That means the two right triangles are in fact 30-60-90s:

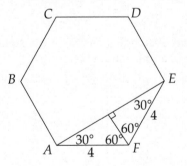

The hypotenuse of each 30–60–90 is 4, so the shared short leg is 2, and the long legs are each $2\sqrt{3}$. Segment $\overline{AE}$ is composed of two legs of $2\sqrt{3}$, so $AE = 4\sqrt{3} \approx 6.93$.

31. C

Never intersecting means parallel or, in other words, having the same slope. The slope of the line having the equation in the question stem is 4. Put the answer choices into slope-intercept form until you find the one with a slope of 4:

(A) $y + 4x = 3$

$y = -4x + 3$ Slope $= -4$

(B) $2y + 4x = 2$

$2y = -4x + 2$

$y = -2x + 1$ Slope $= -2$

(C) $y - 4x = -3$

$y = 4x - 3$ Slope $= 4$

32. C

Use FOIL:

$$(3 + i)(3 - i) = 3 \cdot 3 + 3(-i) + i(3) - i^2$$
$$= 9 - 3i + 3i - i^2$$
$$= 9 - i^2$$
$$= 9 - (-1) = 10$$

33. D

The two right triangles are congruent 3-4-5s.

Sine is opposite over hypotenuse, so $\sin x = \dfrac{3}{5}$, and I is true. Tangent is opposite over adjacent, so $\tan y = \dfrac{4}{3}$ and $\tan r = \dfrac{3}{4}$, and II is not true. Cosine is adjacent over hypotenuse and sine is opposite over hypotenuse, so $\cos x = \dfrac{4}{5}$ and $\sin s = \dfrac{4}{5}$, and III is true.

34. E

To find the volume of the sphere, you need the radius, which is the same as the radius of the semicircle. Find the x-intercepts—that is, the points at which $y = 0$:

$$\sqrt{6x - x^2} = 0$$
$$6x - x^2 = 0$$
$$x(6 - x) = 0$$
$$x = 0 \text{ or } 6$$

The distance between the x-intercept points is the diameter, so diameter $= 6$ and radius $= 3$. Now plug $r = 3$ into the sphere volume formula:

Volume of sphere $= \dfrac{4}{3}\pi r^3 = \dfrac{4}{3}\pi(3)^3 = 36\pi$

35. A

To perform the function on x means to double x. To perform the function twice would mean to double x and then to double the result—in effect turning x into $4x$. For some particular value of x, when you perform the function twice, you end up with $x + 1$. Therefore, for this value of x, $4x$ and $x + 1$ are the same:

$$4x = x + 1$$
$$3x = 1$$
$$x = \dfrac{1}{3}$$

36. D

Because ℓ_1 and ℓ_2 are parallel, you will need to look for transversals. Think of the two angles marked $x°$ as one angle measuring $2x°$. That angle and the angle marked $(4x-30)°$ are exterior angles on the same side of the transversal, so they add up to $180°$:

$$2x + (4x - 30) = 180$$
$$6x - 30 = 180$$
$$6x = 210$$
$$x = 35$$

37. E

To get the total number of possibilities, multiply: 3 colors, 2 body types, and 3 engine types means $3 \times 2 \times 3 = 18$ different models.

38. C

Solve the inequalities separately, then find their intersection:

$$|2x| < 4$$
$$-4 < 2x < 4$$
$$-2 < x < 2$$

$$|x - 2| < 2$$
$$-2 < x - 2 < 2$$
$$0 < x < 4$$

The intersection of these ranges is $0 < x < 2$.

39. B

When the polynomial you're looking for is divided by $3x + 1$, the result is $x^2 - x + 3$ with a remainder of 2. To reconstruct the original polynomial, multiply $3x + 1$ by $x^2 - x + 3$ and add 2:

$$f(x) = (3x+1)(x^2-x+3)+2$$
$$= 3x^3 - 3x^2 + 9x + x^2 - x + 3 + 2$$
$$= 3x^3 - 2x^2 + 8x + 5$$

40. B

Mark up the figure. Fill in the angles you know:

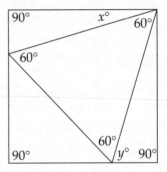

The right triangle at the top and the one on the right are congruent—they have the same hypotenuse and the same longer leg—so the two unknown angles in the upper right are equal:

$$2x + 60 = 90$$
$$2x = 30$$
$$x = 15$$

And now that you know the other two angles in that triangle with y, you can solve for y:

$$y + 15 + 90 = 180$$
$$y = 75$$

Therefore, $y - x = 75 - 15 = 60$.

41. A

Think of the original statement as "If p, then q," in which p is "x is a member of set S," and q is "x is not a member of set T." "If p, then q" is equivalent to "If not q, then not p," or "If x is a member of set T, then x is not a member of set S." That's statement I. Statement II does not follow because it's "If q, then p," and statement III does not follow because it's "If not p, then not q."

42. D

The perimeter of triangle ABC is 20. One side is given as 4, so the other two sides add up to 16. The three sides cannot be 4, 4, and 12, because that violates the Triangle Inequality Theorem. The sides must be 4, 8, and 8:

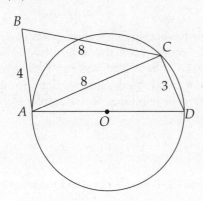

You know that triangle ACD is a right triangle because side AD is a diameter and C is a point on the circle. The legs of triangle ACD are 8 and 3, so the hypotenuse $AD = \sqrt{8^2 + 3^2} = \sqrt{73}$. The radius is half that, or $\dfrac{\sqrt{73}}{2}$. Plug $r = \dfrac{\sqrt{73}}{2}$ into the circle area formula:

$$\text{Area of Circle} = \pi \left(\frac{\sqrt{73}}{2} \right)^2 = \frac{73\pi}{4} \approx 57.33$$

43. A

Sketch a diagram:

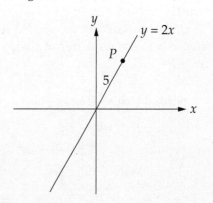

The coordinates of point P are the legs of the right triangle of hypotenuse 5:

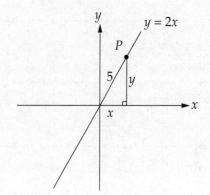

So $x^2 + y^2 = 5^2$. Furthermore, $y = 2x$, so

$$x^2 + (2x)^2 = 25$$
$$5x^2 = 25$$
$$x^2 = 5$$
$$x = \sqrt{5} = 2.24$$
$$y = 2x = 2\sqrt{5} \approx 4.47$$

44. B

The radius of circle O is 4, so the area of circle O is $\pi(4)^2 = 16\pi$. The radius of circle P is 2, so the area of circle P is $\pi(2)^2 = 4\pi$. Think of the shaded region as the difference between a quarter of circle O and a half of circle P:

$$\text{Shaded area} = \frac{1}{4}(16\pi) - \frac{1}{2}(4\pi)$$
$$= 4\pi - 2\pi$$
$$= 2\pi \approx 6.28$$

45. D

Sketch a diagram:

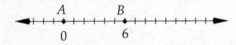

Now, where to put P so that it's twice as far from A as from B. One possibility is two-thirds of the way from A to B, that is, at 4:

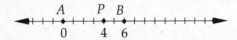

But that's not the only possibility. There's also the point that's the same distance to the right of B as B is from A, that is, at 12:

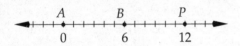

46. B

You're looking for sets of three positive integers with a product of 32. The easiest way to do that is just to list them systematically. If the smallest dimension is 1, then the product of the other two dimensions is 32, so there are these possibilities that include an edge of 1:

$$1 \times 1 \times 32$$
$$1 \times 2 \times 16$$
$$1 \times 4 \times 8$$

If the smallest dimension is 2, then the product of the other two dimensions is 16, so there are these additional possibilities:

$$2 \times 2 \times 8$$
$$2 \times 4 \times 4$$

You can't have a smallest dimension of 4, because then the product of the other two dimensions would have to be 8, and you can't make a product of 8 out of two integers greater than or equal to 4. So the five possibilities listed above are the only possibilities.

47. A

Re-express the equation in standard form:

$$2x^2 + k = 8x$$
$$2x^2 - 8x + k = 0$$

Then use the quadratic formula:

$$x = \frac{-b \pm \sqrt{b^2 - 4ac}}{2a}$$
$$= \frac{8 \pm \sqrt{64 - 8k}}{4}$$

This expression will be real as long as what's under the radical is nonnegative:

$$64 - 8k \geq 0$$
$$-8k \geq -64$$
$$k \leq 8$$

48. A

First put the inequality into standard form:

$$x^2 + 6 < 5x$$
$$x^2 - 5x + 6 < 0$$

Then factor:

$$(x - 2)(x - 3) < 0$$

This product will be negative only when the smaller factor $x - 3$ is negative and the larger factor $x - 2$ is positive. That's when $2 < x < 3$.

49. C

Sketch a diagram:

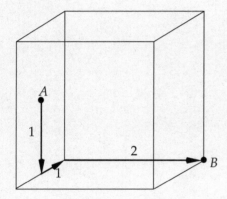

To get from A to B, you go down 1, back 1, and over 2. The straight-line distance, then, is

$$d = \sqrt{1^2 + 1^2 + 2^2} = \sqrt{6} \approx 2.45$$

50. E

First find 5 percent of $87\frac{1}{2}$:

$$(0.5)(87.5) = 4.375$$

An average daily gain of $\frac{1}{2}$ for four days means a net gain of 2. To go up 5 percent for the week, the stock needs to go up another $4.375 - 2 = 2.375$, which is the same as $2\frac{3}{8}$.

HOW TO CALCULATE YOUR SCORE

Step 1: Figure out your raw score. Use the answer key to count the number of questions you answered correctly and the number of questions you answered incorrectly. (Do not count any questions you left blank.) Multiply the number wrong by 0.25 and subtract the result from the number correct. Round the result to the nearest whole number. This is your raw score.

SAT Subject Test: Mathematics Level 1 — Practice Test 3

Number right	Number wrong	Raw score

$$\Box \ - \left(0.25 \times \Box\right) \ = \ \Box$$

Step 2: Find your scaled score. In the Score Conversion Table below, find your raw score (rounded to the nearest whole number) in one of the columns to the left. The score directly to the right of that number will be your scaled score.

A note on your practice test scores: Don't take these scores too literally. Practice test conditions cannot precisely mirror real test conditions. Your actual SAT Subject Test: Mathematics Level 1 score will almost certainly vary from your practice test scores. However, your scores on the practice tests will give you a rough idea of your range on the actual exam.

Conversion Table

Raw	Scaled	Raw	Scaled	Raw	Scaled	Raw	Scaled
50	800	34	610	18	460	2	320
49	790	33	600	17	450	1	310
48	780	32	590	16	450	0	310
47	770	31	580	15	440	−1	300
46	750	30	570	14	430	−2	290
45	740	29	560	13	420	−3	280
44	730	28	550	12	400	−4	270
43	720	27	540	11	390	−5	260
42	700	26	530	10	390	−6	260
41	690	25	520	9	380	−7	250
40	680	24	510	8	370	−8	240
39	670	23	500	7	360	−9	230
38	660	22	500	6	350	−10	230
37	650	21	490	5	350	−11	220
36	640	20	480	4	340	−12	210
35	620	19	470	3	330		

Answer Grid
Practice Test 3

1. Ⓐ Ⓑ Ⓒ Ⓓ Ⓔ
2. Ⓐ Ⓑ Ⓒ Ⓓ Ⓔ
3. Ⓐ Ⓑ Ⓒ Ⓓ Ⓔ
4. Ⓐ Ⓑ Ⓒ Ⓓ Ⓔ
5. Ⓐ Ⓑ Ⓒ Ⓓ Ⓔ
6. Ⓐ Ⓑ Ⓒ Ⓓ Ⓔ
7. Ⓐ Ⓑ Ⓒ Ⓓ Ⓔ
8. Ⓐ Ⓑ Ⓒ Ⓓ Ⓔ
9. Ⓐ Ⓑ Ⓒ Ⓓ Ⓔ
10. Ⓐ Ⓑ Ⓒ Ⓓ Ⓔ
11. Ⓐ Ⓑ Ⓒ Ⓓ Ⓔ
12. Ⓐ Ⓑ Ⓒ Ⓓ Ⓔ
13. Ⓐ Ⓑ Ⓒ Ⓓ Ⓔ
14. Ⓐ Ⓑ Ⓒ Ⓓ Ⓔ
15. Ⓐ Ⓑ Ⓒ Ⓓ Ⓔ
16. Ⓐ Ⓑ Ⓒ Ⓓ Ⓔ
17. Ⓐ Ⓑ Ⓒ Ⓓ Ⓔ
18. Ⓐ Ⓑ Ⓒ Ⓓ Ⓔ
19. Ⓐ Ⓑ Ⓒ Ⓓ Ⓔ
20. Ⓐ Ⓑ Ⓒ Ⓓ Ⓔ
21. Ⓐ Ⓑ Ⓒ Ⓓ Ⓔ
22. Ⓐ Ⓑ Ⓒ Ⓓ Ⓔ
23. Ⓐ Ⓑ Ⓒ Ⓓ Ⓔ
24. Ⓐ Ⓑ Ⓒ Ⓓ Ⓔ
25. Ⓐ Ⓑ Ⓒ Ⓓ Ⓔ

26. Ⓐ Ⓑ Ⓒ Ⓓ Ⓔ
27. Ⓐ Ⓑ Ⓒ Ⓓ Ⓔ
28. Ⓐ Ⓑ Ⓒ Ⓓ Ⓔ
29. Ⓐ Ⓑ Ⓒ Ⓓ Ⓔ
30. Ⓐ Ⓑ Ⓒ Ⓓ Ⓔ
31. Ⓐ Ⓑ Ⓒ Ⓓ Ⓔ
32. Ⓐ Ⓑ Ⓒ Ⓓ Ⓔ
33. Ⓐ Ⓑ Ⓒ Ⓓ Ⓔ
34. Ⓐ Ⓑ Ⓒ Ⓓ Ⓔ
35. Ⓐ Ⓑ Ⓒ Ⓓ Ⓔ
36. Ⓐ Ⓑ Ⓒ Ⓓ Ⓔ
37. Ⓐ Ⓑ Ⓒ Ⓓ Ⓔ
38. Ⓐ Ⓑ Ⓒ Ⓓ Ⓔ
39. Ⓐ Ⓑ Ⓒ Ⓓ Ⓔ
40. Ⓐ Ⓑ Ⓒ Ⓓ Ⓔ
41. Ⓐ Ⓑ Ⓒ Ⓓ Ⓔ
42. Ⓐ Ⓑ Ⓒ Ⓓ Ⓔ
43. Ⓐ Ⓑ Ⓒ Ⓓ Ⓔ
44. Ⓐ Ⓑ Ⓒ Ⓓ Ⓔ
45. Ⓐ Ⓑ Ⓒ Ⓓ Ⓔ
46. Ⓐ Ⓑ Ⓒ Ⓓ Ⓔ
47. Ⓐ Ⓑ Ⓒ Ⓓ Ⓔ
48. Ⓐ Ⓑ Ⓒ Ⓓ Ⓔ
49. Ⓐ Ⓑ Ⓒ Ⓓ Ⓔ
50. Ⓐ Ⓑ Ⓒ Ⓓ Ⓔ

right

wrong

Use the answer key following the test to count up the number of questions you got right and the number you got wrong. (Remember not to count omitted questions as wrong.) "How to Calculate Your Score" on the previous page will show you how to find your score.

Practice Test 3

50 Questions (1 hour)

Directions: For each question, choose the BEST answer from the choices given. If the precise answer is not among the choices, choose the one that best approximates the answer. Then fill in the corresponding oval on the answer sheet.

Notes:

(1) To answer some of these questions, you will need a calculator. You must use at least a scientific calculator, but programmable and graphing calculators are also allowed.

(2) All angle measures on this test are in degrees, so your calculator should be set to degree mode.

(3) Figures in this test are drawn as accurately as possible UNLESS it is stated in a specific question that the figure is not drawn to scale. All figures are assumed to lie in a plane unless otherwise specified.

(4) The domain of any function f is assumed to be the set of all real numbers x for which $f(x)$ is a real number, unless otherwise indicated.

Reference Information: Use the following formulas as needed.

Right circular cone: If r = radius and h = height, then Volume = $\frac{1}{3}\pi r^2 h$, and if c = circumference of the base and ℓ = slant height, then Lateral Area = $\frac{1}{2}c\ell$.

Sphere: If r = radius, then Volume = $\frac{4}{3}\pi r^3$ and Surface Area = $4\pi r^2$.

Pyramid: If B = area of the base and h = height, then Volume = $\frac{1}{3}Bh$.

1.   If $2^{3x-2} = 16$, then $x =$

(A) $\dfrac{1}{2}$ (B) 1   (C) 2   (D) $\dfrac{3}{2}$ (E) 3

2.   If $4a^2 - 4b = 5$ and $a = \dfrac{1}{2}$, then $b =$

(A)   −2
(B)   −1
(C)   0
(D)   1
(E)   2

3.   If $\dfrac{2}{3x + 12} = \dfrac{2}{3}$, then $x + 4 =$

(A) $\dfrac{1}{2}$ (B) 1   (C) $\dfrac{3}{2}$ (D) 2   (E) 3

4.   Which of the following ordered pairs is the solution to the equations $2y - 4x = 4$ and $y + x + 1 = 0$?

(A)   (1, 1)
(B)   (1, 0)
(C)   (0, −1)
(D)   (0, 1)
(E)   (−1, 0)

GO ON TO THE NEXT PAGE

KAPLAN

5. If Sam types at a rate of x words per minute, how many minutes, in terms of x, will it take him to type 500 words?

DO YOUR FIGURING HERE.

(A) $500x$

(B) $500 - x$

(C) $500 + x$

(D) $\dfrac{500}{x}$

(E) $\dfrac{x}{500}$

6. Mary spends 40% of her monthly earnings on rent and 10% of the remaining amount on entertainment. If she spends $800 for rent, how much does she use for entertainment?

(A) $120
(B) $180
(C) $200
(D) $800
(E) $2,000

7. In right triangle ABC, $\angle C$ is a right angle, $AB = 15$, and $BC = 7$. What is the measure of $\angle B$, to the nearest degree?

(A) $25°$
(B) $28°$
(C) $62°$
(D) $65°$
(E) $82°$

8. If $\dfrac{x-3}{6} - \dfrac{x+1}{4} = 3$, $x =$

(A) -45 (B) -15 (C) 1 (D) 15 (E) 45

GO ON TO THE NEXT PAGE

9.   If $f(x) = x + 3$ and $g(x) = f(5x)$, $g(2) =$

DO YOUR FIGURING HERE.

(A) 1   (B) 5   (C) 10   (D) 13   (E) 23

10.   Which of the following lines has the same y-intercept as $y = 3x + 1$ and is perpendicular to $y = \frac{1}{4}x + 2$?

(A) $y = -\frac{1}{3}x + 1$

(B) $y = 4x + 1$

(C) $y = -3x + 2$

(D) $y = -4x + 1$

(E) $y = 3x + 2$

11.   In Figure 1, if triangle ABC is reflected over line l, what will be the coordinates of the reflection of point B?

(A)   (4, 1)
(B)   (4, 4)
(C)   (4, 6)
(D)   (4, 7)
(E)   (4, 8)

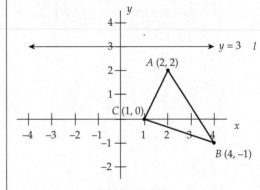

Figure 1

12.   Coach Hathaway is arranging 9 players in a batting order that will include all 9 of his players. How many different arrangements are possible?

(A)   18
(B)   72
(C)   81
(D)   181,440
(E)   362,880

13.   Dick and Suzanne drove for 13 hours at 55 miles per hour. If they had traveled at a rate of 65 miles per hour, how much time, in hours, would they have saved?

(A) 0.5   (B) 1   (C) 2   (D) 6   (E) 11

GO ON TO THE NEXT PAGE

KAPLAN

14. If x and y are both positive odd integers, which of the following must be an odd integer?

   (A) $x + y$

   (B) $5x + 3y$

   (C) $\dfrac{x}{y}$

   (D) $x - y$

   (E) xy

DO YOUR FIGURING HERE.

15. In Figure 2, square $JKLM$ is inscribed in circle O. If the radius is 6, what is the area of the shaded region, to the nearest tenth?

   (A) 10.3   (B) 18.2   (C) 22.8   (D) 28.3   (E) 38.6

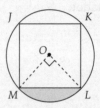

Figure 2

16. If $x = 6$ when $y = 5$ and x varies directly as y, what is the value of x when $y = 2$?

   (A) 1.6$\overline{7}$ (B) 2.4   (C) 7.5   (D) 15   (E) 60

17. $\left| \, |{-5}| - 16 + |{-1}| \, \right| =$

   (A) –12   (B) –10   (C) 10   (D) 12   (E) 22

18. Rob rented a minivan at A's Auto for $60.00 per day with tax included, plus $0.25 per mile. If he rented it for 3 days and was charged $206.00, how many miles did he drive the minivan?

   (A) 6.5   (B) 26   (C) 104   (D) 584   (E) 824

19. If $f(x) = 3x^{-2}$, $f(2) =$

   (A) $\dfrac{1}{36}$ (B) $\dfrac{1}{4}$ (C) $\dfrac{3}{4}$ (D) 12   (E) 36

GO ON TO THE NEXT PAGE

KAPLAN

20.   If $i^2 = -1$, what is the value of $3i^2 + i^3 - i^4$?

    (A) $-4 - i$
    (B) $-2 - i$
    (C) $2 + i$
    (D) $4 + i$
    (E) $6 + 2i$

DO YOUR FIGURING HERE.

21.   An operation $\&$ is defined for all real numbers c and d by the equation $c \& d = \dfrac{c}{4} - \dfrac{d+1}{5}$. If $6 \& d = 1.7$, what is the value of d ?

    (A) -5 (B) -2 (C) 0   (D) 3   (E) 5

22.   If each side of an equilateral triangle is 8, what is the approximate measure of the altitude ?

    (A) 1.73   (B) 2   (C) 3.46   (D) 4   (E) 6.93

23.   For what value of x is $f(x) = 6 + (x - 2)^2$ at its minimum?

    (A) -6 (B) -4 (C) 0   (D) 2   (E) 5

24.   In Figure 3, if $ABCD$ is a parallelogram, $m\angle ADE = 40°$, and $m\angle C = 110°$, what is $m\angle BED$?

    (A) $30°$
    (B) $40°$
    (C) $70°$
    (D) $110°$
    (E) $150°$

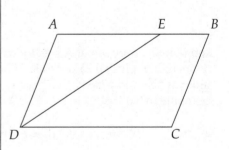

Figure 3

Note: Figure not drawn to scale.

25.   A 3-digit code is made up of three different digits from the set {2, 3, 4, 5, 6}. If 4 is always the first digit in the code, how many 3-digit codes can be formed using each digit only once?

    (A) 3   (B) 5   (C) 12   (D) 13   (E) 20

GO ON TO THE NEXT PAGE

KAPLAN

26. $\sqrt{\dfrac{9\sin^2\theta + 9\cos^2\theta}{4}} =$

(A)   1.5
(B)   2.25
(C)   8.9
(D)   23.1
(E)   56.12

DO YOUR FIGURING HERE.

27. X and Y are independent events. If the probability event X will occur is 0.2 and the probability event Y will occur is 0.9, what is the probability both X and Y will occur?

(A) 18%   (B) 20%   (C) 50%   (D) 72%   (E) 90%

28. If $x^2 - 4x - 12 = 0$, what is the sum of the two possible values of x ?

(A) −8   (B) −6   (C) −2   (D) 2   (E) 4

29. The number of employees who work at Company A increased by 18% during the year 2003. During 2004, the number of employees increased by 26%. What was the approximate percent of increase, to the nearest percent, in employment over the two-year period?

(A) 31%   (B) 44%   (C) 49%   (D) 53%   (E) 65%

30. If 2 tons of snow fall in 1 minute, how many tons of snow fall in 3 hours?

(A)   6
(B)   36
(C)   360
(D) 3,600
(E) 4,000

GO ON TO THE NEXT PAGE

31.   A line m with a slope of $\dfrac{1}{5}$ passes through the points
      $(-4,1)$ and $(1,y)$. What is the value of y ?

      (A) -2 (B) -1 (C) 0   (D) 1   (E) 2

DO YOUR FIGURING HERE.

32.   The mean of a set of 5 numbers is 90. If one of the
      numbers is removed, the mean of the remaining numbers
      is 92. What number was removed?

      (A) 82   (B) 84   (C) 87   (D) 90   (E) 92

33.   Which statement is true for the set of numbers
      $\{6, 7, 11, 6, 8\}$?

      (A)   mean > mode
      (B)   median > mean
      (C)   mode > median
      (D)   mean = median
      (E)   median = mode

34.   A number is randomly selected from the set $\{6, 7, 8, 8, 8,$
      $10, 10, 11\}$. What is the probability the number will be less
      than the mean?

      (A) $\dfrac{1}{4}$ (B) $\dfrac{3}{8}$ (C) $\dfrac{5}{8}$ (D) $\dfrac{3}{4}$ (E) 1

35.   If $3(x + 5) - (x + 2) = 2x - 3x + 4$, then $x =$

      (A) -3 (B) -1 (C) 0   (D) 1   (E) 3

GO ON TO THE NEXT PAGE

KAPLAN

DO YOUR FIGURING HERE.

36. For all $x \neq 0$, $\dfrac{5}{\left(\dfrac{1}{x^2}\right)} =$

(A) $5x^{-2}$

(B) $5x^2$

(C) $\dfrac{x^2}{5}$

(D) $\dfrac{5}{x^2}$

(E) $25x^2$

37. $(x + y + 3)(x + y + 3) =$

(A) $(x + y)^2 + 9$

(B) $x^2 + y^2 + 9$

(C) $x^2 + y^2 + 6xy + 9$

(D) $(x + y)^2 + 6(x + y) + 9$

(E) $(x + y)^2 + 9(x + y) + 9$

38. The area of square $ABCD$ is three-fourths the area of parallelogram $EFGH$. The area of parallelogram $EFGH$ is one-third the area of trapezoid $IJKL$. If square $ABCD$ has an area of 125 square feet, what is the area of trapezoid $IJKL$, in square feet?

(A) 75   (B) 225   (C) 350   (D) 500   (E) 625

39. The circle in Figure 4 has center O, minor arc $AB = 5\pi$, and $m\angle AOB = 50°$. What is the area of circle O ?

(A) 25π
(B) 324π
(C) 576π
(D) $1,296\pi$
(E) $2,500\pi$

Figure 4

Note: Figure not drawn to scale.

GO ON TO THE NEXT PAGE

40.   In Figure 5, *ABCD* is a trapezoid in which $\overline{AB}$ is parallel to $\overline{CD}$, $\overline{AD}$ is perpendicular to $\overline{DC}$, and $\overline{AD}$ is perpendicular to $\overline{AB}$. What is the length of $\overline{BC}$?

(A) 3   (B) 4   (C) 5   (D) 6   (E) 7

41.   If the square root of the cube root of *x* is 3, what is the value of *x* ?

(A)   3
(B)   9
(C)   27
(D)   729
(E)   531,441

42.   The line $y = -1$ intersects the parabola $y = x^2 - 10x + 24$ at one point. What is the vertex of the parabola?

(A)   (–5, –1)
(B)   (0, –1)
(C)   (5, –1)
(D)   (24, –1)
(E)   (35, –1)

43.   In Figure 6, if the length of $\overline{BC}$ equals one-half the length of $\overline{AB}$, the sum of *BC* and *CD* equals three-fourths the length of $\overline{AD}$, and the length of $\overline{AD}$ equals 16, what is the distance between point *C* and the midpoint of $\overline{AD}$?

(A) 2   (B) 4   (C) 6   (D) 8   (E) 10

44.   If $12^5 = 3^t \times 4^t$, what is the value of *t* ?

(A) 2.5   (B) 5   (C) 10   (D) 20   (E) 40

DO YOUR FIGURING HERE.

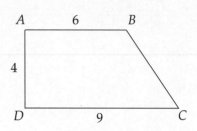

Figure 5

Note: Figure not drawn to scale.

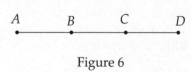

Figure 6

Note: Figure not drawn to scale.

GO ON TO THE NEXT PAGE

45. For what values of x will $|2x - 1| > 5$?

    (A) $-3 < x < 2$

    (B) $x < 3$

    (C) $x > -2$

    (D) $-2 < x < 3$

    (E) $x < -2$ or $x > 3$

46. Which of the following has the greatest value?

    (A) $(6^2 \times 6)^5$

    (B) $(36)^5$

    (C) $(36^2 \times 6^3)^2$

    (D) $(216)^4$

    (E) $(6^4)^4$

47. If $5xy = 2$, what is the value of $(5xy)^{5xy}$?

    (A) 4   (B) 10   (C) 25   (D) 100   (E) 525

48. In Figure 7, $\overline{ST}$ and $\overline{QR}$ are parallel, $\overline{RU}$ and $\overline{QT}$ are perpendicular, $m\angle RQU = x + 4$, and $m\angle QTS = 2x - 30$. What is the $m\angle QRU$?

    (A) $34°$

    (B) $38°$

    (C) $52°$

    (D) $68°$

    (E) $88°$

DO YOUR FIGURING HERE.

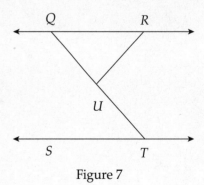

Figure 7

GO ON TO THE NEXT PAGE

KAPLAN

49. In the right circular cylinder shown in Figure 8, points E and F are the centers of the bases, and segment AB is the diameter of one of the bases. What is the volume of the cylinder, to the nearest integer, if $AE = 10$ and $m\angle AEF = 30°$?

(A)   393
(B)   680
(C)   785
(D)   2,721
(E)   3,142

50. In Figure 9, $AB = 9$, $BC = 5$, and $m\angle C = 50°$. What is the area of triangle ABC, to the nearest tenth?

(A)   21.7
(B)   22.5
(C)   36.4
(D)   40.2
(E)   43.5

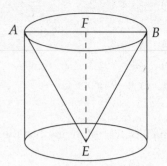

Figure 8

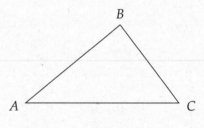

Figure 9

STOP!

If you finish before time is up, you may check your work.

**Turn the page for
answers and explanations
to Practice Test 3.**

Answer Key

1. C	18. C	35. A
2. B	19. C	36. B
3. B	20. A	37. D
4. E	21. B	38. D
5. D	22. E	39. B
6. A	23. D	40. C
7. C	24. E	41. D
8. A	25. C	42. C
9. D	26. A	43. A
10. D	27. A	44. B
11. D	28. E	45. E
12. E	29. C	46. E
13. C	30. C	47. A
14. E	31. E	48. C
15. A	32. A	49. B
16. B	33. A	50. A
17. C	34. C	

ANSWERS AND EXPLANATIONS

1. C

Express each side as a power of the same base.

$$2^{3x-2} = 16$$
$$2^{3x-2} = 2^4$$
$$3x - 2 = 4$$
$$3x = 6$$
$$x = 2$$

2. B

Use substitution.

$$4a^2 - 4b = 5$$
$$4\left(\frac{1}{2}\right)^2 - 4b = 5$$
$$4\left(\frac{1}{4}\right) - 4b = 5$$
$$1 - 4b = 5$$
$$-4b = 4$$
$$b = -1$$

3. B

You don't have to solve for the value of x to answer this question because the denominator on the left side of the equation is 3 times $(x + 4)$. Noticing this can save you time for other questions—a good strategy in the early questions.

$$\frac{2}{3x + 12} = \frac{2}{3}$$
$$\frac{2}{3(x + 4)} = \frac{2}{3}$$

Since $3(1) = 3$, $x + 4 = 1$.

4. E

First isolate one of the variables:

$$y + x + 1 = 0$$
$$y = -x - 1$$

Then use substitution:

$$2y - 4x = 4$$
$$2(-x - 1) - 4x = 4$$
$$-2x - 2 - 4x = 4$$
$$-6x - 2 = 4$$
$$-6x = 6$$
$$x = -1$$

Plug $x = -1$ back into one of the original equations to find y:

$$y + x + 1 = 0$$
$$y - 1 + 1 = 0$$
$$y = 0$$

That's (E).

5. D

If y is the number of minutes it takes to type 500 words, then set up the following proportion:

$$\frac{x \text{ words}}{1 \text{ minute}} = \frac{500 \text{ words}}{y \text{ minutes}}$$
$$xy = 500$$
$$y = \frac{500}{x}$$

6. A

Mary spends 40% of her earnings on rent.

If x = the amount earned, $x(40\%) = 800$

$$0.4x = 800$$
$$x = \$2,000 \text{ monthly earnings}$$

Mary has 60% of her earnings left after spending 40% on rent. She spends 10% of 60% on entertainment which is 10% × 60% = 6% on entertainment.

6% of $2,000 = 0.06 × 2,000 = \$120

KAPLAN

7. C

Make a diagram. Use cosine to find $m\angle B$.
(Remember: SOHCAHTOA)

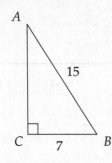

$$\cos\angle B = \frac{\text{adjacent}}{\text{hypotenuse}}$$

$$\cos\angle B = \frac{7}{15}$$

$$m\angle B = 62°$$

8. A

First, rewrite the left side of the equation as a single fraction by getting a common denominator.

$$\frac{x-3}{6} - \frac{x+1}{4} = 3$$

$$\frac{2(x-3)}{12} - \frac{3(x+1)}{12} = 3$$

$$\frac{2x-6-3x-3}{12} = 3$$

$$\frac{-x-9}{12} = 3$$

Multiplying both sides by 12, you get

$$-x - 9 = 36$$

$$-x = 45$$

$$x = -45$$

9. D

Since $g(x) = f(5x)$, $g(2) = f(5(2)) = f(10)$.

Since $f(x) = x + 3$, $f(10) = 10 + 3 = 13$.

10. D

For any equation in the form $y = mx + b$, m = slope and b = y-intercept. For the line to have the same y-intercept as $y = 3x + 1$, the y-intercept must be 1. Perpendicular lines have slopes that are negative reciprocals of each other.

The negative reciprocal of $\frac{1}{4}$ is -4.

11. D

The reflection of a point or figure is the mirror image of that point or figure. When point B is reflected over the line $y = 3$, the x-coordinate will not change. Point B is 4 units below the line $y = 3$, so it will become 4 units above the line.

12. E

This is the number of permutations of 9 things taken 9 at a time.

$$_9P_9 = 9! = 9 \times 8 \times 7 \times 6 \times 5 \times 4 \times 3 \times 2 \times 1 = 362,880$$

13. C

Distance = rate × time.

If they traveled at a rate of 55 m/hr:
$d = 55$ m/hr × 13 hrs = 715 miles.

Now we know they traveled 715 miles. If they traveled at a rate of 65 m/hr:

$$715 = 65 \text{ m/hr} \times x \text{ hours}$$

$$x = 11 \text{ hours}$$

Finally, 13 hours – 11 hours = 2 hours of time saved.

14. E

(A) is always even because two odd integers added together always produce an even integer. (B) is always even because the sum of 5 times an odd number and 3 times an odd number must be even. (C) is not an integer unless x = a whole multiple of y, so it is not *always* an odd integer. (D) is always even because the difference of two odd integers must be an even number. (E), however, is always odd because the product of two odd integers is odd.

Note: Another way to solve this problem is to pick odd numbers such as 3 and 5 for x and y.

15. A

Cut the circle into 4 equal slices. One of those slices will include the shaded region and a right triangle formed by two radii and the chord. To find the area of the shaded region, subtract the area of the triangle from the area of one-fourth of the circle.

Area of one-fourth of a circle $= \frac{1}{4}\pi r^2 = \frac{1}{4}(36)\pi = 9\pi$.

Area of the triangle $= \frac{1}{2}bh = \frac{1}{2}(6)(6) = 18$.

Area of the shaded region $= 9\pi - 18 \approx 10.3$.

16. B

If x varies directly as y, $\frac{x}{y} = k$.

$$\frac{6}{5} = k$$

$$\frac{6}{5} = \frac{x}{2}$$

$$\frac{12}{5} = x$$

$$x = 2.4$$

17. C

$||-5|-16+|-1|| = |5 - 16 + 1| = |-10| = 10$.

18. C

Since Rob rented the minivan for 3 days, he will pay $3(\$60) = \$180.00$ plus 0.25 times the number of miles he drove it.

Let m = the number of miles driven. Then $206 = 180 + 0.25m$, $26 = 0.25m$, and $m = 104$ miles.

19. C

$$f(2) = 3(2)^{-2} = 3\left[\frac{1}{2}\right]^2 = 3\left[\frac{1}{4}\right] = \frac{3}{4}.$$

20. A

$$3i^2 + i^3 - i^4 = 3i^2 + i^2 \times i - i^2 \times i^2$$
$$= 3(-1) + (-1)i - (-1)(-1)$$
$$= -3 - i - 1$$
$$= -4 - i$$

Alternatively, you can factor out i^2 from the equation to get $i^2(3 + i - i^2)$. Next substitute -1 for i^2 in the new equation: $-1(3 + i - (-1)) = -1(3 + i + 1) = -1(4 + i)$. Then distribute -1: $-4 - i$.

21. B

$$6 \& d = \frac{6}{4} - \frac{d+1}{5} = 1.7$$

$$\frac{3}{2} - \frac{d+1}{5} = 1.7$$

Get a common denominator on the left side:
$$\frac{15}{10} - \frac{2(d+1)}{10} = 1.7 .$$

$$\frac{15 - 2d - 2}{10} = 1.7$$

$$\frac{13 - 2d}{10} = \frac{1.7}{1}$$

Cross multiply: $13 - 2d = 17$, so $-2d = 4$, and $d = -2$.

22. E

Use the Pythagorean theorem:

$$8^2 = 4^2 + b^2$$
$$64 = 16 + b^2$$
$$48 = b^2$$
$$\sqrt{48} = b$$
$$b \approx 6.93$$

Alternatively, draw an equilateral triangle including the altitude. Two congruent 30–60–90 triangles are formed. The sides in a 30–60–90 are in a ratio of $1 : \sqrt{3} : 2$.

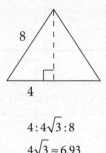

$$4 : 4\sqrt{3} : 8$$

$$4\sqrt{3} \approx 6.93$$

23. D

The expression $6 + (x - 2)^2$ is at its minimum when $(x - 2)^2$ is at its minimum value. Since the expression is squared, its minimum is zero. $(x - 2)^2 = 0$ when $x = 2$.

24. E

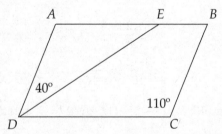

If $m\angle C = 110°$, then the $m\angle A = 110°$ since opposite angles of a parallelogram are equal. Since the three angles in a triangle add up to $180°$, $m\angle AED = 30°$. $\angle AED$ and $\angle BED$ are supplementary, so $m\angle BED = 150°$.

25. C

The first digit can only be the number 4. Therefore, there is only one option for this digit. There are 4 digits left to choose from for the second digit and 3 digits to choose for the third. Multiply to get 12, which is the number of 3-digit codes that can be formed using each digit only once.

Alternatively, the first digit has already been determined, so look at all of the options for the last two digits: 23, 25, 26, 32, 35, 36, 52, 53, 56, 62, 63, or 65.

26. A

Use the trigonometric identity $sin^2\theta + cos^2\theta = 1$.

$$\sqrt{\frac{9\,sin^2\theta + 9\,cos^2\theta}{4}} = \sqrt{\frac{9(sin^2\theta + cos^2\theta)}{4}}$$

$$\sqrt{\frac{9(1)}{4}} = \sqrt{\frac{9}{4}} = \frac{3}{2} = 1.5$$

27. A

When two events are independent, the probability they will both occur is the product of each of their probabilities.

$$0.2\,(0.9) = 0.18 = 18\%$$

28. E

This answer can be found by finding both roots and adding them together.

$$x^2 - 4x - 12 = 0$$
$$(x - 6)(x + 2) = 0$$
$$x - 6 = 0 \text{ or } x + 2 = 0$$
$$x = 6 \text{ or } x = -2$$
$$6 + (-2) = 4$$

29. C

Let P = the number of employees at the beginning of 2003. By the end of 2003, the number of employees increases by 18% of P or $0.18P$, so the number becomes $P + 0.18P = 1.18P$.

At the beginning of 2004, the number of employees is represented by $1.18P$. By the end of 2004, the number of employees increases by 26% of $1.18P$, so the number becomes

$$1.18P + 0.26(1.18P) \approx 1.18P + 0.31P = 1.49P.$$

Since $1.49P = P + 0.49P$, the increase over the two-year period is approximately 49%.

30. C

Set up a proportion.

$$\frac{2 \text{ tons}}{1 \text{ minute}} = \frac{x \text{ tons}}{3 \text{ hours}}$$

$$\frac{2 \text{ tons}}{1 \text{ minute}} = \frac{x \text{ tons}}{180 \text{ minutes}}$$

$$1x = 2 \times 180$$

$$x = 360$$

31. E

$$\text{slope} = \frac{y_2 - y_1}{x_2 - x_1}$$

$$\frac{1}{5} = \frac{y - 1}{1 + 4}$$

$$\frac{1}{5} = \frac{y - 1}{5}$$

Set the numerators equal since the denominators are equal.

$$1 = y - 1$$

$$y = 2$$

32. A

Since the mean of a set of 5 numbers is 90, the sum of those numbers is $5 \times 90 = 450$. When one number is removed, the mean of the set of the 4 remaining numbers is 92. The sum of those numbers is $4 \times 92 = 368$. Then $450 - 368 = 82$, so 82 must have been the number that was removed.

33. A

Mean (average) $= \dfrac{6 + 7 + 11 + 6 + 8}{5} = \dfrac{38}{5} = 7.6$.

Median = 7. (Be sure to place the numbers in order before selecting the middle number.)

Mode = 6. (Occurs the most often in the set.)

(A) is the only statement that is true; mean > mode.

34. C

Mean (average) =
$$\frac{6 + 7 + 8 + 8 + 8 + 10 + 10 + 11}{8} = \frac{68}{8} = 8.5.$$

Out of eight numbers, five are less than the mean of 8.5.

35. A

$$3(x + 5) - (x + 2) = 2x - 3x + 4$$

$$3x + 15 - x - 2 = -x + 4$$

$$2x + 13 = -x + 4$$

$$3x = -9$$

$$x = -3$$

36. B

$$\frac{5}{\left(\dfrac{1}{x^2}\right)} = 5 \div \left(\frac{1}{x^2}\right) = 5 \times x^2 = 5x^2$$

37. D

It helps to put parentheses around $x + y$. Then use FOIL.

$$[(x + y) + 3][(x + y) + 3] =$$
$$(x + y)^2 + 3(x + y) + 3(x + y) + 9 =$$
$$(x + y)^2 + 6(x + y) + 9$$

38. D

Area of the square $= \dfrac{3}{4}$ (Area of the parallelogram).

Area of the parallelogram $= \dfrac{1}{3}$ (Area of the trapezoid).

Using substitution, Area of the square $= \dfrac{3}{4} \times \dfrac{1}{3}$ (Area of the trapezoid).

Area of the square $= \dfrac{1}{4}$ (Area of the trapezoid).

Thus, $125 = \dfrac{1}{4}$ (Area of the trapezoid).

Area of the trapezoid $= 4 \times 125 = 500$.

39. B

Find the circumference first in order to find the radius.

Set up a proportion.

$$\frac{50}{360} = \frac{5\pi}{C}$$
$$50C = 360(5\pi)$$
$$C = 36\pi$$
$$C = \pi d$$
$$d = 36$$
$$r = 18$$
$$A = \pi r^2 = \pi(18)^2 = 324\pi$$

40. C

Draw the altitude from point B. The sides of the right triangle form a 3-4-5 Pythagorean triple. $BC = 5$.

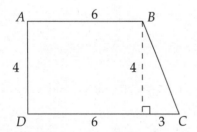

41. D

The square root of the cube root of x can be expressed as $\sqrt{\sqrt[3]{x}} = x^{\frac{1}{3}\left(\frac{1}{2}\right)} = x^{\frac{1}{6}}$.

$$x^{\frac{1}{6}} = 3$$
$$x = 3^6 = 729$$

42. C

The point where the horizontal line and the parabola intersect must be the vertex of the parabola. In order to find the point of intersection, set the equations equal to each other.

$$-1 = x^2 - 10x + 24$$
$$0 = x^2 - 10x + 25$$
$$0 = (x - 5)(x - 5)$$
$$x = 5$$

Since $x = 5$ and the vertex is on the line $y = -1$, the coordinates of the vertex are $(5, -1)$.

43. A

$$BC = \frac{1}{2}AB$$

$$BC + CD = \frac{3}{4}AD$$

Since $AD = 16$, $BC + CD = \frac{3}{4}(16) = 12$.

Since $BC + CD = 12$, $AB = 4$.

Since $BC = \frac{1}{2}AB$,

$$BC = \frac{1}{2}(4) = 2.$$

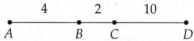

The midpoint of $\overline{AD}$ is 8 units from point A. Point C is 6 units from point A. The distance between them is $8 - 6 = 2$.

44. B

$$12^5 = 3^t \times 4^t$$

$$12^5 = 12^t$$

Since the bases are the same, set the exponents equal to each other.

$$t = 5$$

45. E

$|2x - 1| > 5$ means that $2x - 1$ is more than 5 units from zero.

$$2x - 1 > 5, 2x > 6, x > 3$$

or

$$2x - 1 < -5, 2x < -4, x < -2$$

46. E

(A) $(6^2 \times 6)^5 = (6^3)^5 = 6^{15}$.

(B) $(36)^5 = (6^2)^5 = 6^{10}$.

(C) $(36^2 \times 6^3)^2 = (6^4 \times 6^3)^2 = (6^7)^2 = 6^{14}$.

(D) $(216)^4 = (6^3)^4 = 6^{12}$.

(E) $(6^4)^4 = 6^{16}$.

47. A

When $5xy = 2$, $(5xy)^{5xy} = 2^2 = 4$.

48. C

Since alternate interior angles of parallel lines are congruent, $m\angle RQU = m\angle QTS$.

$$x + 4 = 2x - 30$$

$$x = 34°$$

$$m\angle RQU = 34 + 4 = 38°$$

$m\angle QRU = 180 - (90 + 38) = 180 - 128 = 52°$.

49. B

Since triangle AEF is a 30-60-90 triangle, the lengths of the sides are in a ratio of $1:\sqrt{3}:2$. $AF = \dfrac{1}{2} AE$, and $EF = AF\sqrt{3}$.

$$AF = \frac{1}{2} \times 10 = 5$$
$$EF = 5\sqrt{3} \approx 8.66$$

Volume $= \pi r^2 h = \pi(5^2)8.66 \approx 680$.

50. A

In order to use the formula $A = \dfrac{1}{2}bh$, find AC and the length of the height from vertex B.

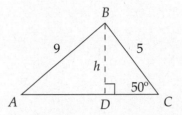

$$\sin 50° = \frac{h}{5}$$

Use a calculator to determine the value of h.

$$h = 3.83$$

Use the Pythagorean theorem to find the length of $\overline{AD}$ and $\overline{DC}$.

$$3.83^2 + AD^2 = 9^2$$

$$AD = 8.14$$

$$3.83^2 + CD^2 = 5^2$$

$$CD = 3.21$$

$$AC = AD + DC = 11.35$$

$$A = \frac{1}{2}(11.35)(3.83) = 21.7$$

HOW TO CALCULATE YOUR SCORE

Step 1: Figure out your raw score. Use the answer key to count the number of questions you answered correctly and the number of questions you answered incorrectly. (Do not count any questions you left blank.) Multiply the number wrong by 0.25 and subtract the result from the number correct. Round the result to the nearest whole number. This is your raw score.

SAT Subject Test: Mathematics Level 1 — Practice Test 4

Number right		Number wrong		Raw score
☐	− (0.25 ×	☐	) =	☐

Step 2: Find your scaled score. In the Score Conversion Table below, find your raw score (rounded to the nearest whole number) in one of the columns to the left. The score directly to the right of that number will be your scaled score.

A note on your practice test scores: Don't take these scores too literally. Practice test conditions cannot precisely mirror real test conditions. Your actual SAT Subject Test: Mathematics Level 1 score will almost certainly vary from your practice test scores. However, your scores on the practice tests will give you a rough idea of your range on the actual exam.

Conversion Table

Raw	Scaled	Raw	Scaled	Raw	Scaled	Raw	Scaled
50	800	34	610	18	460	2	320
49	790	33	600	17	450	1	310
48	780	32	590	16	450	0	310
47	770	31	580	15	440	−1	300
46	750	30	570	14	430	−2	290
45	740	29	560	13	420	−3	280
44	730	28	550	12	400	−4	270
43	720	27	540	11	390	−5	260
42	700	26	530	10	390	−6	260
41	690	25	520	9	380	−7	250
40	680	24	510	8	370	−8	240
39	670	23	500	7	360	−9	230
38	660	22	500	6	350	−10	230
37	650	21	490	5	350	−11	220
36	640	20	480	4	340	−12	210
35	620	19	470	3	330		

Answer Grid
Practice Test 4

1. (A) (B) (C) (D) (E)
2. (A) (B) (C) (D) (E)
3. (A) (B) (C) (D) (E)
4. (A) (B) (C) (D) (E)
5. (A) (B) (C) (D) (E)
6. (A) (B) (C) (D) (E)
7. (A) (B) (C) (D) (E)
8. (A) (B) (C) (D) (E)
9. (A) (B) (C) (D) (E)
10. (A) (B) (C) (D) (E)
11. (A) (B) (C) (D) (E)
12. (A) (B) (C) (D) (E)
13. (A) (B) (C) (D) (E)
14. (A) (B) (C) (D) (E)
15. (A) (B) (C) (D) (E)
16. (A) (B) (C) (D) (E)
17. (A) (B) (C) (D) (E)
18. (A) (B) (C) (D) (E)
19. (A) (B) (C) (D) (E)
20. (A) (B) (C) (D) (E)
21. (A) (B) (C) (D) (E)
22. (A) (B) (C) (D) (E)
23. (A) (B) (C) (D) (E)
24. (A) (B) (C) (D) (E)
25. (A) (B) (C) (D) (E)

26. (A) (B) (C) (D) (E)
27. (A) (B) (C) (D) (E)
28. (A) (B) (C) (D) (E)
29. (A) (B) (C) (D) (E)
30. (A) (B) (C) (D) (E)
31. (A) (B) (C) (D) (E)
32. (A) (B) (C) (D) (E)
33. (A) (B) (C) (D) (E)
34. (A) (B) (C) (D) (E)
35. (A) (B) (C) (D) (E)
36. (A) (B) (C) (D) (E)
37. (A) (B) (C) (D) (E)
38. (A) (B) (C) (D) (E)
39. (A) (B) (C) (D) (E)
40. (A) (B) (C) (D) (E)
41. (A) (B) (C) (D) (E)
42. (A) (B) (C) (D) (E)
43. (A) (B) (C) (D) (E)
44. (A) (B) (C) (D) (E)
45. (A) (B) (C) (D) (E)
46. (A) (B) (C) (D) (E)
47. (A) (B) (C) (D) (E)
48. (A) (B) (C) (D) (E)
49. (A) (B) (C) (D) (E)
50. (A) (B) (C) (D) (E)

right

wrong

Use the answer key following the test to count up the number of questions you got right and the number you got wrong. (Remember not to count omitted questions as wrong.) "How to Calculate Your Score" on the previous page will show you how to find your score.

Practice Test 4

50 Questions (1 hour)

Directions: For each question, choose the BEST answer from the choices given. If the precise answer is not among the choices, choose the one that best approximates the answer. Then fill in the corresponding oval on the answer sheet.

Notes:

(1) To answer some of these questions, you will need a calculator. You must use at least a scientific calculator, but programmable and graphing calculators are also allowed.

(2) All angle measures on this test are in degrees, so your calculator should be set to degree mode.

(3) Figures in this test are drawn as accurately as possible UNLESS it is stated in a specific question that the figure is not drawn to scale. All figures are assumed to lie in a plane unless otherwise specified.

(4) The domain of any function f is assumed to be the set of all real numbers x for which $f(x)$ is a real number, unless otherwise indicated.

Reference Information: Use the following formulas as needed.

Right circular cone: If r = radius and h = height, then Volume = $\frac{1}{3}\pi r^2 h$, and if c = circumference of the base and ℓ = slant height, then Lateral Area = $\frac{1}{2}c\ell$.

Sphere: If r = radius, then Volume = $\frac{4}{3}\pi r^3$ and Surface Area = $4\pi r^2$.

Pyramid: If B = area of the base and h = height, then Volume = $\frac{1}{3}Bh$.

---

1. What is a possible form of the quadratic equation, if the sum of the roots is –1 and the product of the roots is –20?

    (A) $y = 2x^2 + 2x - 40$
    (B) $y = x^2 - x - 20$
    (C) $y = x^2 + x + 20$
    (D) $y = 2x^2 - 2x - 20$
    (E) $-20y = x - 1$

2. A boat sights the top of a 40-foot lighthouse at an angle of elevation of 25 degrees. How far away is the boat from the lighthouse (the horizontal distance), to the nearest tenth of a foot?

    (A)   16.9 feet
    (B)   18.7 feet
    (C)   44.1 feet
    (D)   85.8 feet
    (E)   94.6 feet

3. Figure 1 shows a square with total area of 121 square units. Based on the figure, what is the value of x?

    (A)   2
    (B)   4
    (C)   7
    (D)   11
    (E)   49

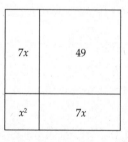

Figure 1

4. A plumbing company charges $42.50 plus x dollars per hour to make a house call. Which expression below represents the cost of a 90-minute house call?

    (A) $90(42.50 + x)$
    (B) $1.5(42.50 + x)$
    (C) $1.5(42.50x)$
    (D) $42.50 + 90x$
    (E) $42.50 + 1.5x$

GO ON TO THE NEXT PAGE

5.   Solve for x: $64^x = 4^{x^2-4}$.

(A) $x = 4$ or $x = -1$
(B) $x = -4$ or $x = 1$
(C) $x = 10$
(D) $x = \sqrt{20}$
(E) $x = 3$

6.   A rectangular solid has three faces with areas of 28, 20, and 35 square centimeters. What is the volume of this solid?

(A)   83 cubic cm
(B)   140 cubic cm
(C)   166 cubic cm
(D)   196 cubic cm
(E)   19,600 cubic cm

7.   Solve for x, to the nearest tenth: $\log_{12} 640 = x$.

(A)   1.7
(B)   2.6
(C)   2.8
(D)   53.3
(E)   7,680

8.   A triangle with coordinates (2,0), (2,6), and (6,0) is rotated about the line $x = 2$. What is the volume, in cubic units, of the resultant solid?

(A) 16   (B) 8π   (C) 16π   (D) 32π   (E) 48π

GO ON TO THE NEXT PAGE

KAPLAN

9. Within a given area code, how many seven-digit numbers are available, given the restrictions that the first three digits cannot be 911 or 411 and the first digit cannot be a 1 or a 0?

    (A)   604,800
    (B)   2,893,401
    (C)   6,380,000
    (D)   7,980,000
    (E)   8,000,000

DO YOUR FIGURING HERE.

10. Solve for x: $x^2 + 7x \geq 30$.

    (A) $-3 \leq x \leq 10$
    (B) $-10 \leq x \leq 3$
    (C) $x \leq -10$ or $x \geq 3$
    (D) $x \leq -3$ or $x \geq 10$
    (E) $-11 \leq x \leq 4$

11. What is the point of intersection of the line that passes through the points $(1,5)$ and $(3,9)$ and the line perpendicular to that line that passes through the point $(0,-2)$?

    (A) $(-2,-1)$
    (B) $(-2,2)$
    (C) $(2,-2)$
    (D) $(-1,-2)$
    (E) $(1,2)$

12. Solve for x: $|\, 4x - 3 \,| - 9 < 0$.

    (A) $1.5 \leq x \leq 3$
    (B) $x < 3$
    (C) $x < 1.5$ or $x > 3$
    (D) $x \leq -3$
    (E) $-1.5 < x < 3$

GO ON TO THE NEXT PAGE

KAPLAN

13. Given that $i = \sqrt{-1}$, what is the multiplicative inverse of $5 - i$?

(A) $5 + i$

(B) $\dfrac{5 + i}{26}$

(C) $\dfrac{1}{5 + i}$

(D) $\dfrac{5 + i}{24}$

(E) $\dfrac{5 - i}{24}$

14. What is the value of x: $\log_x 125 = 3$?

(A)   5
(B)   41.7
(C)   122
(D)   128
(E)   375

15. Given $f(x) = 7x + 5$, what is $f^{-1}(x)$, the inverse function?

(A) $f^{-1}(x) = 7x + 5$

(B) $f^{-1}(x) = 5x + 7$

(C) $f^{-1}(x) = \dfrac{1}{7}x + \dfrac{1}{5}$

(D) $f^{-1}(x) = \dfrac{1}{7}x - 5$

(E) $f^{-1}(x) = \dfrac{x - 5}{7}$

DO YOUR FIGURING HERE.

GO ON TO THE NEXT PAGE

KAPLAN

16. Take 6 less than a number n. If you raise this result to the 5th power, it is equal to 32. What is the value of n?

    (A) 2
    (B) 2.07
    (C) 4 only
    (D) 4 or 8
    (E) 8 only

DO YOUR FIGURING HERE.

17. Take a fraction that is equivalent to $\frac{3}{5}$ and add 2 to both the numerator and the denominator. If the result is $\frac{5}{8}$, what was the original fraction?

    (A) $\frac{8}{14}$

    (B) $\frac{3}{6}$

    (C) $\frac{13}{22}$

    (D) $\frac{18}{30}$

    (E) Cannot be determined.

GO ON TO THE NEXT PAGE

18.   Simplify: $\left(x - 3 - \dfrac{4}{x} \right) \left(\dfrac{3}{x+1} \right)$.

DO YOUR FIGURING HERE.

(A) $\dfrac{3(x-4)}{x}$

(B) $2x - 4$

(C) $\dfrac{-6(x+2)}{x+1}$

(D) -7

(E) -12

19.   When $3x^3 + 2x^2 + 2x + k$ is divided by $x + 2$, the remainder is 4. What is the value of k ?

(A)   4
(B)   24
(C)   34
(D)   54
(E) $x + 4$

20.   Simplify: $\dfrac{(8x^2 y^3)^3 z^2}{4x^{-2} y^5 z^4}$.

(A) $2x^4 y z^2$

(B) $\dfrac{128x^8 y^4}{z^2}$

(C) $2x^8 y^4 z^2$

(D) $\dfrac{128 x^8 y}{z^2}$

(E) $6y^4 z^2$

GO ON TO THE NEXT PAGE

KAPLAN

21.   What is the solution to this system of equations?

$$y = x^2 + 4x - 1$$

$$-3x + y = 1$$

(A)   (1,4) and (−2,5)
(B)   (1,−4) and (2,11)
(C)   (1,4) and (−2,−5)
(D)   (−2,7)
(E)   (0,1)

22.   Which of the following graphs is NOT a function?

(A)

(B)

(C)

(D)

(E)

GO ON TO THE NEXT PAGE

23.   What is the solution to this system of equations?

$$x^2 + y^2 = 25$$

$$x - y = 5$$

(A)   (–5,0) only
(B)   (0,5) only
(C)   (0,–5) and (–5,0)
(D)   (0,5) and (5,0)
(E)   (0,–5) and (5,0)

24.   A bus travels from point *A* to point *B*. The bus starts out going slowly and then drives faster. Traffic is then at a total standstill due to an accident. The bus continues at the faster speed. The bus reaches point *B*. Which of the following graphs could represent the bus's distance from point *A*, as a function of time?

(A)

(B)

(C)

(D)

(E)

GO ON TO THE NEXT PAGE

KAPLAN

25. A parallelogram has angles of measure 45 degrees and 135 degrees. The shorter side of the parallelogram measures 2.83 meters, and the other side is 8.83 meters. What is the area of the parallelogram, to the nearest hundredth?

(A)   8.83 m²

(B)   12.49 m²

(C)   17.67 m²

(D)   23.32 m²

(E)   24.99 m²

26. A fitness club offers weights, treadmills, and aerobic classes. Fifty members were asked which services they used, and the results were as follows:

18 members use weights and do aerobics.

25 members use weights and the treadmill.

15 members use the treadmill and do aerobics.

10 members use all three activities.

How many members use only one of the three services?

(A) 0   (B) 10   (C) 12   (D) 18   (E) 40

27. Given the equation $y = \dfrac{x + 7a}{x - a^2}$, what is x in terms of y and a ?

(A) $x = \dfrac{y + 7a}{y - a^2}$

(B) $x = \dfrac{y + 7a}{ya^2}$

(C) $x = \dfrac{ya^2 + 7a}{y - 1}$

(D) $x = y(a^2 + 7a)$

(E) $x = (y + 7a)(y - a^2)$

GO ON TO THE NEXT PAGE

DO YOUR FIGURING HERE.

KAPLAN

28. Solve for x to the nearest hundredth: $\sqrt[3]{\dfrac{2x+3}{5}} = \dfrac{2}{3}$.

    (A)   −0.76
    (B)   −0.69
    (C)   −0.67
    (D)   0.69
    (E)   0.76

29. If $x = y^3 + 4y$ and $y = \dfrac{7}{k}$, what is the approximate value of x when $k = 21$?

    (A) 1.37   (B) 1.44   (C) 1.67   (D) 4.04   (E) 4.11

30. Solve for x: $\dfrac{1}{x} + \dfrac{4}{5x} = \dfrac{2}{x+5}$.

    (A)   0.71
    (B)   3.57
    (C)   5.8
    (D)   25
    (E)   45

31. A car travels a steady speed of x miles per hour. How many hours, in terms of x, will it take to drive 540 miles?

    (A) $540x$

    (B) $540 + x$

    (C) $\dfrac{x}{540}$

    (D) $\dfrac{540}{x}$

    (E) $x - 540$

GO ON TO THE NEXT PAGE

KAPLAN

32.   In the segment shown in Figure 2, the length of $\overline{BC}$ is three more than twice the length of $\overline{AB}$. If the length of $\overline{AC}$ is 27 cm, what is the length of $\overline{AB}$?

(A)   8 cm
(B)   9 cm
(C)   10 cm
(D)   12 cm
(E)   15 cm

DO YOUR FIGURING HERE.

Figure 2

33.   If $f(x) = x^2 - 10$ and $g(x) = 4x + 3$, what is $f(g(2))$?

(A) –24   (B) –21   (C) 12   (D) 27   (E) 111

34.   Which graph below represents the solution to the following system of inequalities?

$$y < x + 2$$

$$y > -3x + 3$$

(A)

(B)

(C)

(D)

(E)

GO ON TO THE NEXT PAGE

35. In the rectangular solid shown in Figure 3, what is the length of diagonal $\overline{BE}$, to the nearest tenth?

(A)   7.07
(B)   8.6
(C)   10
(D)   13.60
(E)   25

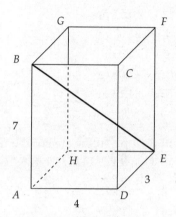

Figure 3

36. In the segment shown in Figure 4, the ratio of the lengths of $\overline{AB}$ to $\overline{AC}$ is 5:8. If x represents AB, what is the midpoint of $\overline{AC}$ in terms of x?

(A) $4x$

(B) $\dfrac{4x}{5}$

(C) $\dfrac{8x}{5}$

(D) $\dfrac{5x}{2}$

(E) $\dfrac{5x}{16}$

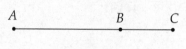

Figure 4

37. Let ☉ be defined as $a☉b = a^2 + ba - 16b \div a$. What is the value of $8☉3$?

(A)   −2.67
(B)   5
(C)   34
(D)   46
(E)   82

GO ON TO THE NEXT PAGE

38. In Figure 5, a circle is inscribed in a square. What is the area of the shaded portion, to the nearest hundredth?

   (A) 10.52 m²
   (B) 21.03 m²
   (C) 22.6 m²
   (D) 42.06 m²
   (E) 84.13 m²

39. In Figure 6, if the area of the circle is 64π square units, what is the area of triangle *ABC* to the nearest hundredth?

   (A) 55.43
   (B) 110.85
   (C) 128
   (D) 221.7
   (E) 443.41

40. In Figure 7, a regular hexagon of side length 5 cm is inscribed in a circle. What percentage of the circle is shaded, to the nearest tenth?

   (A) 13.6%
   (B) 17.3%
   (C) 78.5%
   (D) 82.7%
   (E) 86.4%

41. A dog is chained on a 6-foot leash, fastened to the corner of a rectangular building. About how much area does the dog have to move in?

   (A) 27 ft²
   (B) 36 ft²
   (C) 56.55 ft²
   (D) 84.82 ft²
   (E) 113.10 ft²

DO YOUR FIGURING HERE.

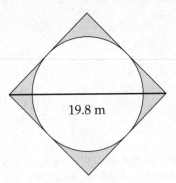

19.8 m

Figure 5

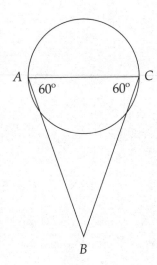

A   60°   60°   *C*

B

Figure 6

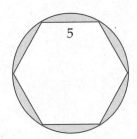

5

Figure 7

GO ON TO THE NEXT PAGE

KAPLAN

42. What is the domain of the function $f(x) = \dfrac{1}{\sqrt{x^2 - 16}}$?

(A) All the real numbers

(B) $x < -4$ or $x > 4$

(C) $x \geq 4$

(D) $x > 8$

(E) $x < -8$ or $x > 8$

43. Given the following stem-and-leaf plot and the statements shown below the plot, which of the statements are true?

The number of customers in line for an attraction

```
2 | 2 2 2 2 4 7
3 | 1 2 2 3 5 6 7
4 | 5 7 7
5 | 0 1 1 1
```

where 3 | 6 means 36 customers

I. The mode is equal to the median.

II. The median is less than the mean.

III. The mean is 33.

(A) I only

(B) II only

(C) III only

(D) I and II only

(E) II and III only

GO ON TO THE NEXT PAGE

Use the cards shown below for questions 44 and 45

$$\boxed{M}\boxed{A}\boxed{T}\boxed{H}\boxed{E}\boxed{M}\boxed{A}\boxed{T}\boxed{I}\boxed{C}\boxed{S}$$

DO YOUR FIGURING HERE.

44. What is the probability of picking an M card at random, without replacement, and then an A card, at random, without looking?

  (A) $\dfrac{4}{11}$

  (B) $\dfrac{4}{21}$

  (C) $\dfrac{4}{22}$

  (D) $\dfrac{2}{55}$

  (E) $\dfrac{4}{121}$

45. What is the probability of picking an S card at random, without replacement, and then NOT picking a T card, at random, without looking?

  (A) $\dfrac{4}{55}$

  (B) $\dfrac{9}{121}$

  (C) $\dfrac{2}{11}$

  (D) $\dfrac{10}{22}$

  (E) $\dfrac{10}{11}$

GO ON TO THE NEXT PAGE

46. In right triangle *EDF* in Figure 8, the length of $\overline{DF}$ is 2 cm, and the length of $\overline{EF}$ is 7 cm. What is the measure of ∠*EFD*, to the nearest hundredth of a degree?

    (A)   15.95°
    (B)   16.6°
    (C)   73.40°
    (D)   90°
    (E)   99.9°

47. If $f(x) = x^2 - 7$, then $f(a - 3)$ is

    (A) $a^2 - 6a - 16$
    (B) $a^2 - 10$
    (C) $a^2 + 21$
    (D) $a^2 - 6a + 2$
    (E) $2a - 13$

48. Given point *A*(−3,−8), if the midpoint of segment *AB* is (1,−5), what are the coordinates of point *B* ?

    (A)   (5,−2)
    (B)   (4,−2)
    (C)   (−1,−6.5)
    (D)   (−2,−2)
    (E)   (−1,−1.5)

49. The area of the triangle with coordinates (1,2), (5,5), and (*k*,2) is 15 square units. What is a possible value for *k* ?

    (A)   −10
    (B)   −9
    (C)   −5
    (D)   5
    (E)   6

Figure 8

GO ON TO THE NEXT PAGE

KAPLAN

50.   Given right triangle *RST* in Figure 9, what is the length of
      $\overline{ST}$, to the nearest hundredth?

      (A)   12.04 mm
      (B)   13.38 mm
      (C)   16.21 mm
      (D)   24.22 mm
      (E)   26.90 mm

DO YOUR FIGURING HERE.

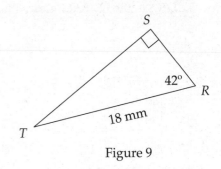

Figure 9

STOP!

**If you finish before time is up,
you may check your work.**

**Turn the page for
answers and explanations
to Practice Test 4.**

Answer Key

1. A	18. A	35. B
2. D	19. B	36. B
3. B	20. B	37. E
4. E	21. C	38. D
5. A	22. D	39. B
6. B	23. E	40. B
7. B	24. A	41. D
8. D	25. C	42. B
9. D	26. C	43. B
10. C	27. C	44. D
11. A	28. A	45. A
12. E	29. A	46. C
13. B	30. E	47. D
14. A	31. D	48. A
15. E	32. A	49. B
16. E	33. E	50. A
17. D	34. C	

ANSWERS AND EXPLANATIONS

1. A

First, eliminate choice (E), since it is a linear equation, not a quadratic equation. Then you could attempt to factor each equation to determine the roots, then check the sum and the product of these roots. However, you may recall that in a quadratic equation in the form $ax^2 + bx + c$, the sum of the roots is $\frac{-b}{a}$, and the product of the roots is $\frac{c}{a}$. In choice (A), $\frac{-b}{a} = \frac{-2}{2} = -1$, and $\frac{c}{a} = \frac{-40}{2} = -20$.

2. D

Draw a picture of the problem situation:

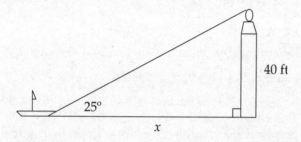

As you can see from the drawing, this is a right triangle, where x represents the horizontal distance of the boat to the lighthouse. Use the tangent function to find the distance. The tangent of the angle is equal to the ratio of the lengths of $\frac{\text{opposite side}}{\text{adjacent side}}$. Set up the equation $\tan(25°) = \frac{40}{x}$. Multiply both sides by x and then divide both sides by $\tan(25°)$ to get $x = \frac{40}{\tan(25°)} = 85.8$, to the nearest tenth. Use your calculator to determine this, making sure it's in degree mode.

3. B

The figure shows the geometric equivalent of a perfect square, broken up into a quadratic with pieces $x^2 + 7x + 7x + 49 = 121$, the given area. Factor the quadratic to get $(x + 7)(x + 7) = 121$, or $(x + 7)^2 = 121$. Take the square root of both sides of the equation to get $x + 7 = \pm 11$. Eliminate the root -11, since the length of the side of a square cannot be negative. Subtract 7 from both sides of this equation to get $x = 4$.

4. E

The company charges a flat fee of $42.50, plus x dollars times the number of hours. The problem asks for the cost of a 90-minute call, which is 1.5 hours. The correct expression is therefore $42.50 + 1.5x$.

5. A

To solve this equation, express both sides as powers of the same base, in this case 4: $(4^3)^x = 4^{x^2-4}$. Use the power rule for exponents to get $4^{3x} = 4^{x^2-4}$. Since both sides of the equation are powers of 4, simplify the equation to be $3x = x^2 - 4$, or $x^2 - 3x - 4 = 0$. Factor the left-hand side to get $(x - 4)(x + 1) = 0$. Therefore, either $x = 4$ or $x = -1$. Alternatively, you could arrive at the correct answer by substituting in all the answer choices to find the correct solutions.

6. B

To solve this problem, sketch an "unfolded" drawing of this solid:

	7		
5	35		
20	28	20	4
	35	5	
	28	4	
	7		

Opposite faces of a rectangular solid are congruent, so there are six faces, with the corresponding areas shown above. Look for factors that will produce the given areas. These factors are the dimensions of the solid: $7 \times 4 = 28$, $7 \times 5 = 35$, and $4 \times 5 = 20$. The dimensions are 7, 4, and 5. Use the volume formula to arrive at the correct answer: $V = lwh$, so $V = 7 \times 5 \times 4 = 140$ cubic centimeters.

7. B

You can solve this equation by rewriting it in exponential form: $12^x = 640$. At this point, you could try each value of x to arrive at the correct answer, rounded. Alternatively, use logarithm rules to solve: $\log(12^x) = \log 640$, and $x \log(12) = \log 640$. Divide both sides by the log12 to get $x = \dfrac{\log 640}{\log 12}$, which is 2.6, to the nearest tenth.

8. D

Draw the triangle on the coordinate plane, as shown.

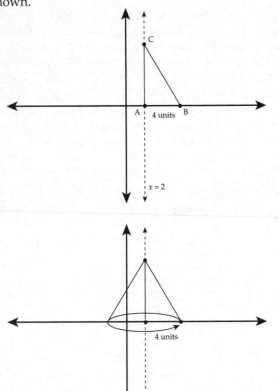

This is a right triangle, with one side on the line $x = 2$. As shown, when this triangle is rotated about $x = 2$, the result is a cone. The radius of the cone is the distance from point A to point B, or $6 - 2 = 4$ units. The height of the cone is the distance from point A to point C, or $6 - 0 = 6$. Use the formula for the volume of a right circular cone, which will be given on the formula sheet: $V = \dfrac{1}{3}\pi r^2 h$, or

$$V = \frac{\pi \times 4^2 \times 6}{3} = 32\pi.$$

9. D

First, determine how many numbers are available with just the restriction on the first digit. For the first digit, there are 8 choices, and then for the

remaining 6 digits, there are 10 choices. So the pool of available numbers is $8 \times 10 \times 10 \times 10 \times 10 \times 10 \times 10$, or 8,000,000. If phone numbers cannot start with the digit pattern 911, then this eliminates $1 \times 1 \times 1 \times 10 \times 10 \times 10 \times 10$ numbers, or 10,000. There are another 10,000 numbers starting with 411. So the actual number of available phone numbers is $8,000,000 - 20,000 = 7,980,000$.

10. C

Subtract 30 from both sides of the inequality to get $x^2 + 7x - 30 \geq 0$. Factor the left-hand side: $(x + 10)(x - 3) \geq 0$. In order for the inequality to be true, either both factors must be positive, or both factors must be negative. This will be true when $x \geq 3$ or $x \leq -10$.

11. A

Determine the equation of the line through the points (1,5) and (3,9). The slope is $= \dfrac{9-5}{3-1} = \dfrac{4}{2}$, which is 2. Find the y-intercept, b: $5 = 2(1) + b$, so $b = 3$. The equation of the line is $y = 2x + 3$. The line perpendicular to this line has a slope that is the negative reciprocal, namely $-\dfrac{1}{2}$, and the y-intercept is -2, because the given point is (0,–2). Now, find the intersection of these lines.

You could graph these lines on a graphing calculator and use the intersect function.

Alternatively, set one equation equal to the other and solve for x. So $2x + 3 = -\dfrac{1}{2}x - 2$. Add $\dfrac{1}{2}x$ and subtract 3 from both sides to get $\dfrac{5}{2}x = -5$. Multiply both sides by $\dfrac{2}{5}$, and $x = -2$. Use this value for x in the first equation to find the corresponding y value. $y = 2(-2) + 3; y = -1$. The intersection point is (–2,–1).

12. E

For an absolute value equation, first isolate the absolute value term by adding 9 to both sides of the equation: $|4x - 3| < 9$. Now, solve two inequalities, $4x - 3 < 9$ and $-(4x - 3) < 9$. The first inequality simplifies to $4x < 12$, or $x < 3$. The second inequality simplifies to $-4x + 3 < 9$, or $-4x < 6$. When you divide by –4, you switch the inequality symbol, so $x > -\dfrac{3}{2}$, or $x > -1.5$. By combining the two inequalities, the solution can be written as $-1.5 < x < 3$.

13. B

The multiplicative inverse is the reciprocal, which is $\dfrac{1}{5-i}$. You must then rationalize the denominator by multiplying by $\dfrac{5+i}{5+i}$:

$$\frac{1}{5-i} \times \frac{5+i}{5+i} = \frac{5+i}{25-i^2}$$

Since $i = \sqrt{-1}$, the inverse becomes

$$\frac{5+i}{25-(-1)} = \frac{5+i}{26}$$

14. A

Rewrite the equation to be $x^3 = 125$. Take the cube root of both sides to get $x = 5$.

15. E

To find the inverse of a function, first write the original equation as $y = 7x + 5$. Now, exchange the x and y values and then solve for y: $x = 7y + 5$. Subtract 5 from both sides of the equation, then divide both sides by 7 to get $f^{-1}(x) = \dfrac{x-5}{7}$.

16. E

Translate the words into algebra. Six less than a number n is $n - 6$. The equation is thus $(n - 6)^5 = 32$. Take the fifth root of each side to get $n - 6 = 2$. Add 6 to both sides of the equation, and $n = 8$.

17. D

A fraction equivalent to $\dfrac{3}{5}$ has the form $\dfrac{3x}{5x}$.

Translate the words to get the equation $\dfrac{3x+2}{5x+2} = \dfrac{5}{8}$.

Solve for x by cross multiplying: $8(3x + 2) = 5(5x + 2)$. Use the distributive property to get $24x + 16 = 25x + 10$. Next, subtract $24x$ and 10 from both sides of the equation to get $x = 6$. Substitute this into the form of the original fraction: $\dfrac{3(6)}{5(6)} = \dfrac{18}{30}$.

18. A

Express the first factor, with the common denominator of x, to get $\dfrac{x^2}{x} - \dfrac{3x}{x} - \dfrac{4}{x} = \dfrac{x^2 - 3x - 4}{x}$.

Factor the numerator: $\dfrac{(x-4)(x+1)}{x}$. When this factor is multiplied by the second, the $(x + 1)$ term will cancel to get $\dfrac{(x-4)}{x} \times \dfrac{3}{1} = \dfrac{3(x-4)}{x}$.

19. B

Divide this polynomial using long division:

$$\begin{array}{r} 3x^2 - 4x + 10 \\ x+2\overline{)3x^3 + 2x^2 + 2x + k} \\ \underline{3x^3 + 6x^2} \\ -4x^2 + 2x \\ \underline{-4x^2 - 8x} \\ 10x + k \\ \underline{10x + 20} \\ k - 20 \end{array}$$

The remainder is given as 4, so $k - 20 = 4$, and $k = 24$.

20. B

First, apply the power rule to the numerator: $(8x^2y^3)^3z^2 = 8^3x^{2\times3}y^{3\times3}z^2$. This simplifies to $512x^6y^9z^2$. Use the division rule of exponents on the numerator and the denominator to get $128x^{6-(-2)}y^{9-5}z^{2-4}$, which simplifies to $\dfrac{128x^8y^4}{z^2}$.

21. C

Solve the second equation for y by adding $3x$ to both sides to get $y = 3x + 1$. Now, set the two equations equal to each other: $3x + 1 = x^2 + 4x - 1$. Subtract $3x$ and 1 from both sides to get $x^2 + x - 2 = 0$. Factor the left-hand side: $(x + 2)(x - 1) = 0$. So $x = -2$, or $x = 1$. Replace x with the value -2 to find the corresponding y-coordinate: $-3(-2) + y = 1$, $6 + y = 1$, $y = -5$. So the first coordinate is $(-2,-5)$. To complete the calculation, replace x with the value 1: $-3(1) + y = 1$, $-3 + y = 1$ $y = 4$. So the other coordinate is $(1,4)$. But note that, since the first coordinate appears in only one choice, you could have stopped after calculating it.

22. D

A function is a relation in which each x value has one unique corresponding y value. In graph D, the relation is not a function. Note the "vertical line test" in graph D below, which indicates that for a value of x, there are two possible y values.

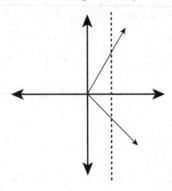

23. E

Solve the second equation for y by adding y and subtracting 5 from both sides to get $y = x - 5$. Now, substitute $x - 5$ for y in the first equation to get $x^2 + (x - 5)^2 = 25$. Multiply to get $x^2 + x^2 - 10x + 25 = 25$. Combine like terms and subtract 25 from both sides to get $2x^2 - 10x = 0$. Factor the left-hand side to get $2x(x - 5) = 0$. So $x = 0$ or $x = 5$. When $x = 0$, $y = 0 - 5 = -5$. When $x = 5$, $y = 5 - 5 = 0$. The points of intersection are $(0,-5)$ and $(5,0)$.

24. A

Graph A accurately describes the story.

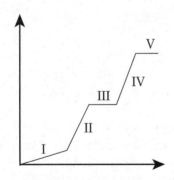

The section marked I indicates the start of the trip. Section II represents the increase in speed; speed is the slope of the line. Section III is the standstill in traffic. Section IV denotes the continuation at the faster speed, and section V is the stop at point B.

25. C

Draw the parallelogram with the height shown.

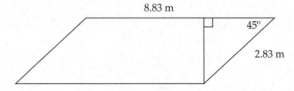

The triangle created inside the parallelogram is an isosceles triangle; two of the sides are congruent. In a 45–45–90 right triangle, if the congruent sides are x units long, then the hypotenuse is $x\sqrt{2}$. The hypotenuse is the shorter side of the parallelogram. It is given that the hypotenuse is 2.83; therefore, the height is $\dfrac{2.83}{\sqrt{2}} \approx 2.0$. The area of a parallelogram is base times the height, so the area = $8.83 \times 2 \approx 17.67$, to the nearest hundredth.

26. C

It helps to draw a Venn diagram of the problem situation:

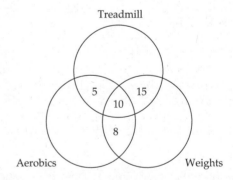

As shown in the diagram, add up the number of members who either do all three activities or any two of the activities: $8 + 10 + 5 + 15 = 38$. The remaining members use only one of the services, so $50 - 38 = 12$ members.

27. C

To solve the equation for x, first multiply both sides by $(x - a^2)$ to get $y(x - a^2) = x + 7a$. Use the distributive property on the left-hand side: $yx - ya^2 = x + 7a$. Add ya^2 and subtract x from both sides of the equation to get $yx - x = ya^2 + 7a$. Factor out the x from the left-hand side: $x(y - 1) = ya^2 + 7a$. Finally, divide both sides by $(y - 1)$ to solve for x in terms of y and a: $\dfrac{ya^2 + 7a}{y - 1}$.

28. A

First, raise both sides of the equation to the third power to eliminate the cube root on the left-hand side. This results in $\dfrac{2x + 3}{5} = \dfrac{2^3}{3^3}$, or $\dfrac{2x + 3}{5} = \dfrac{8}{27}$. Cross multiply to get $27(2x + 3) = 40$, or $54x + 81 = 40$. Subtract 81 from both sides and divide both sides by 54 to get $x = -\dfrac{41}{54} = -0.76$.

29. A

It is given that $k = 21$, so substitute this value in to find the value of y: $y = \dfrac{7}{k} = \dfrac{7}{21} = \dfrac{1}{3}$. Use this value of y in the first equation: $x = y^3 + 4y$, or $x = \left(\dfrac{1}{3}\right)^3 + 4\left(\dfrac{1}{3}\right)$ $= \left(\dfrac{1}{27}\right) + \left(\dfrac{4}{3}\right) = 1.37$.

30. E

Combine the left-hand terms by using a common denominator of $5x$: $\dfrac{5 + 4}{5x} = \dfrac{2}{x + 5}$, or $\dfrac{9}{5x} = \dfrac{2}{x + 5}$. Cross multiply to get $9x + 45 = 10x$. Subtract $9x$ from both sides, and $x = 45$.

31. D

Let h represent the number of hours to drive 540 miles. Set up the proportion of miles over hours: $\dfrac{x}{1} = \dfrac{540}{h}$. Cross multiply to get $xh = 540$. Divide both sides by x to solve for h, the number of hours, to get $h = \dfrac{540}{x}$.

32. A

Let x represent the length of AB. Then translate the words to represent $BC = 2x + 3$. These segments added together equal AC, so $x + 2x + 3 = 27$. Combine like terms and subtract 3 from both sides to get $3x = 24$; divide both sides by 3 to get $x = 8$.

33. E

In a composition of functions, first evaluate the innermost function, in this case $g(2)$: $4 \times 2 + 3 = 8 + 3 = 11$. Now use this result in the outermost function f: $f(11) = 11^2 - 10 = 121 - 10 = 111$.

34. C

(A) and (D) can be immediately eliminated because the boundary lines are solid. The inequalities stated in the problem are "less than" and "greater than," which denote dotted boundary lines. (B) can be eliminated because one of the boundary lines in this graph is a horizontal line. This would be $y < 2$, not $y < x + 2$. (E) is incorrect because the shading would indicate two "less than" inequalities; the shading is below both of the boundary lines.

(C) is the only graph that could represent the system. The first inequality is represented by a dotted line through the y-intercept of 2 with a slope of 1. It is then shaded below this line since the inequality is "less than." The second inequality is represented by a dotted line through the y-intercept of 3 with a slope of -3. It is then shaded above this line since the inequality is "greater than." Where the shadings meet is the solution to the system of inequalities, or graph (C).

35. B

The problem is asking for the length of the diagonal through the rectangular solid. First, determine the length of the diagonal through the bottom base of the solid. This is shown as the bold dotted line (segment AE).

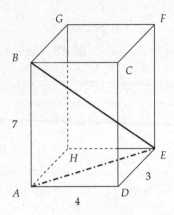

This is the hypotenuse of a right triangle with legs of length 3 and 4. Use the Pythagorean theorem: $a^2 + b^2 = c^2$, or $3^2 + 4^2 = c^2$. This diagonal is therefore $\sqrt{(9+16)} = \sqrt{25} = 5$. (It's also a special 3-4-5 right triangle.) Now, the diagonal of the solid is the hypotenuse of the right triangle with legs of length 5 and 7. In the figure, segment AE is now the leg of a second right triangle, whose hypotenuse is the length to be solved for. Use the Pythagorean theorem: $5^2 + 7^2 = c^2$; $c = \sqrt{(25+49)} = \sqrt{74} \approx 8.6$.

36. B

Because you are given the ratio of AB to AC, first set up a proportion to find the length of AC in terms of x: $\dfrac{5}{8} = \dfrac{x}{AC}$. Cross multiply to get $5(AC) = 8x$, or $AC = \dfrac{8x}{5}$. Divide this result by 2 since the problem asks for the midpoint of $\overline{AC}$, which is therefore $\dfrac{8x}{10} = \dfrac{4x}{5}$.

37. E

Substitute the value of 8 for a and 3 for b to get $8^2 + 8 \times 3 - 16 \times 3 \div 8$. Now, use the order of operations to simplify. First, take 8 squared (64), and then perform multiplication and division from left to right. Multiply to get $64 + 24 - 48 \div 8$. Next perform division: $48 \div 8 = 6$. Then, first add and finally subtract: $64 + 24 - 6 = 82$.

38. D

To find the area of the shaded region, find the area of the square, then subtract out the area of the circle. The square's diagonal length is given as 19.8 m. This is also the hypotenuse of an isosceles right triangle, and the legs of the triangle are the congruent sides of the square. In a 45–45–90 triangle, the hypotenuse is $\sqrt{2}$ times as long as the legs, so the leg length is $\dfrac{19.8}{\sqrt{2}} \approx 14$ m. Since the circle is inscribed in the square, the radius is one-half of the length of a side, or 7 m. Find the area by calculating $A_{square} - A_{circle}$, or $s^2 - \pi r^2$; $14^2 - \pi(7^2)$; $196 - 153.94 = 42.06$.

39. B

Use the given area of the circle, 64π, to find the radius, r: $r^2 = 64$, so $r = 8$. This radius is one-half the base of the equilateral triangle shown, so the base is 16. As shown in the figure below, there are two 30–60–90 right triangles.

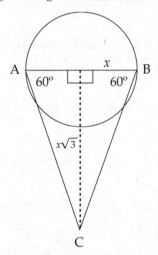

In this type of triangle, the legs are x and $x\sqrt{3}$. Therefore, the height of the triangle is $8\sqrt{3}$ units long. Now the area of the 30–60–90 triangle can be calculated: $A = \dfrac{1}{2} \times 8 \times 8\sqrt{3} \approx 55.43$. The equilateral triangle is two of these; the area is $2 \times 55.425 \approx 110.85$.

40. B

A regular hexagon is made up of six equilateral triangles, in this case with length of 5 units:

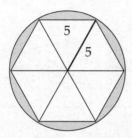

As shown above, the radius is therefore 5 units. The area of the circle is $A = \pi r^2 = 25\pi$. The area of a hexagon with side length x is given as $3 \times x \times \dfrac{x}{2}\sqrt{3}$.

The shaded portion is Area$_{circle}$ – Area$_{hexagon}$, or $25\pi - 3(5)(2.5)(\sqrt{3}) \approx 13.588$. The percentage of the shaded portion can be found by dividing this area by the area of the circle: $13.588 \div 25\pi \approx 0.173$, or 17.3%.

41. D

Draw a picture of the situation to solve the problem:

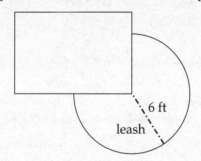

As shown in the diagram above, the dog has $\dfrac{3}{4}$, or 0.75, of a circle in which to move. The circle has a radius equal to the length of the chain, 6 ft. The area the dog has to move in is therefore $0.75(\pi)(6^2) \approx 84.82$ ft^2.

42. B

The domain of this function would be the real numbers, with the exception of any numbers that would produce a zero in the denominator or a negative value within the radicand. The value of 4 will produce a zero in the denominator; any number greater than 4 will be part of the domain. In the same way, because $(-4)^2 = 16$, any number less than -4 will be part of the domain.

43. B

Using the stem-and-leaf plot, calculate the mean, median, and mode to determine the truth of the three statements. The mode is 22, the value that occurs most often. The median is the middle value. In a stem-and-leaf plot, the values are ordered. There are twenty values in the plot, so the median is halfway between the tenth and eleventh value, or halfway between 33 and 35. The median is 34. The mean is the sum of the values divided by 20, or $(22 + 22 + 22 + 22 + 24 + 27 + 31 + 32 + 32 + 33 + 35 + 36 + 37 + 45 + 47 + 47 + 50 + 51 + 51 + 51) \div 20$, or $717 \div 20 = 35.85$. The only true statement is II, that the median (34) is less than the mean (35.85).

44. D

Out of the 11 possible cards, 2 of them are M's. The probability of choosing an M is therefore $\dfrac{2}{11}$. One card is now gone; there are 10 possible cards and 2 of them are A's. This probability of picking an A is $\dfrac{2}{10} = \dfrac{1}{5}$. The probability of this compound event is $\dfrac{2}{11} \times \dfrac{1}{5} = \dfrac{2}{55}$.

45. A

Out of the 11 possible cards, 1 of them is an S. The probability of choosing an S is therefore $\dfrac{1}{11}$. One card is now gone; there are 10 possible cards and 2 of them are T's. So 8 of them are *not* T's. The probability of not drawing a T is $\dfrac{8}{10} = \dfrac{4}{5}$. The probability of this compound event is $\dfrac{1}{11} \times \dfrac{4}{5} = \dfrac{4}{55}$.

46. C

This is a calculator problem. The problem asks for the measure of an acute angle of a right triangle. Use trigonometry to solve. Side $\overline{DF}$ is adjacent to $\angle EFD$, and side $\overline{EF}$ is the hypotenuse. Use the cosine function: $\cos(\angle EFD) = \dfrac{2}{7}$, so the measure of $\angle EFD =$ $\arccos\left(\dfrac{2}{7}\right)$, which is 73.40°, to the nearest hundredth of a degree.

47. D

Substitute the value of $(a - 3)$ for x in the function: $f(a-3) = (a-3)^2 - 7$. Multiply to get $(a-3)(a-3) - 7 = a^2 - 6a + 9 - 7$. Combine like terms: $a^2 - 6a + 2$.

48. A

The coordinates of the midpoint of two points (x_1, y_1) and (x_2, y_2) are $\left(\dfrac{x_1 + x_2}{2}, \dfrac{y_1 + y_2}{2}\right)$. Setting $x_1 = -3$ and $y_1 = -8$, solve for x_2 and y_2. First, $1 = \dfrac{-3 + x_2}{2}$, or $2 = -3 + x_2$. Add 3 to both sides, and $x_2 = 5$. Second, $-5 = \dfrac{-8 + y_2}{2}$, or $-10 = -8 + y_2$. Add 8 to both sides to get $y_2 = -2$.

49. B

Draw the triangle on a coordinate plane, as shown below:

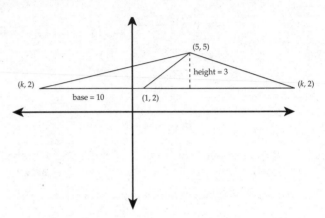

The formula for the area of a triangle is $A = \dfrac{1}{2}bh$. Because the area is given as 15 square units, and the height is $5 - 2 = 3$ units, the base must be 10 units long. Refer to the drawing above and see that the length for the base is the difference between the x-coordinate of $(1,2)$ and the x-coordinate of $(k,2)$. The base can therefore be represented as $k - 1$, or $1 - k$. Therefore, $k = 11$, or $k = -9$.

50. A

The problem shows a right triangle in which an angle measure and the length of the hypotenuse are known. Use trigonometry to find the length of $\overline{ST}$, the opposite side of the 42° angle. Using x to represent the length of $\overline{ST}$, $\sin(42°) = \dfrac{x}{18}$. Multiply both sides by 18, and $x = 18 \times \sin(42°)$. $x = 12.04$ to the nearest hundredth.

100 Essential Math Concepts

The math on the SAT subject tests covers a lot of ground—from arithmetic to algebra to geometry.

Don't let yourself be intimidated. We've highlighted the 100 most important concepts that you'll need for Math 1 and listed them in this chapter.

Use this list to remind yourself of the key areas you'll need to know. Do four concepts a day, and you'll be ready within a month. If a concept continually causes you trouble, circle it and come back to it as you try to do the questions.

You've probably been taught most of these concepts in school already, so this list is a great way to refresh your memory.

NUMBER PROPERTIES

1. Number Categories

Integers are **whole numbers**; they include negative whole numbers and zero.

A **rational number** is a number that can be expressed as a **ratio of two integers**. **Irrational numbers** are real numbers—they have locations on the number line—but they can't be expressed precisely as a fraction or decimal. The most important irrational numbers are $\sqrt{2}$, $\sqrt{3}$, and π.

2. Adding/Subtracting Signed Numbers

To **add a positive and a negative number**, first ignore the signs and find the positive difference between the number parts. Then attach the sign of the original number with the larger number part. For example, to add 23 and −34, first ignore the minus sign and find the positive difference between 23 and 34—that's 11. Then attach the sign of the number with the larger number part—in this case it's the minus sign from the −34. So 23 + (−34) = −11.

Make **subtraction** situations simpler by turning them into addition. For example, you can think of –17 – (–21) as –17 + (+21) or –17 – 21 as –17 + (–21).

To **add or subtract a string of positives and negatives**, first turn everything into addition. Then combine the positives and negatives so that the string is reduced to the sum of a single positive number and a single negative number.

3. Multiplying/Dividing Signed Numbers

To multiply and/or divide positives and negatives, treat the number parts as usual and **attach a minus sign if there were originally an odd number of negatives**. For example, to multiply –2, –3, and –5, first multiply the number parts: $2 \times 3 \times 5 = 30$. Then go back and note that there were *three*—an *odd* number—of negatives, so the product is negative: $(–2) \times (–3) \times (–5) = –30$.

4. PEMDAS

When performing multiple operations, remember to perform them in the right order.

PEMDAS, which means **Parentheses** first, then **Exponents**, then **Multiplication** and **Division** (left to right), and lastly **Addition** and **Subtraction** (left to right). In the expression $9 – 2 \times (5 – 3)^2 + 6 \div 3$, begin with the parentheses: $(5 – 3) = 2$. Then do the exponent: $2^2 = 4$. Now the expression is: $9 – 2 \times 4 + 6 \div 3$. Next do the multiplication and division to get: $9 – 8 + 2$, which equals 3. If you have difficulty remembering PEMDAS, use this sentence to recall it: Please Excuse My Dear Aunt Sally.

5. Counting Consecutive Integers

To count consecutive integers, **subtract the smallest from the largest and add 1**. To count the number of integers from 13 through 31, subtract: $31 – 13 = 18$. Then add 1: $18 + 1 = 19$.

NUMBER OPERATIONS AND CONCEPTS

6. Exponential Growth

If r is the ratio between consecutive terms, a_1 is the first term, a_n is the nth term, and S_n is the sum of the first n terms, then $a_n = a_1 r^{n-1}$ and $S_n = \dfrac{a_1 - a_1 r^n}{1 - r}$.

7. Union and Intersection of Sets

The things in a set are called elements or members. The **union** of Set A and Set B, sometimes expressed as $A \cup B$, is the set of elements that are in either or both of Set A and Set B. If Set $A = \{1, 2\}$ and Set $B = \{3, 4\}$, then $A \cup B = \{1, 2, 3, 4\}$. The **intersection** of Set A and Set B, sometimes expressed as $A \cap B$, is the set of elements common to both Set A and Set B. If Set $A = \{1, 2, 3\}$ and Set $B = \{3, 4, 5\}$, then $A \cap B = \{3\}$.

DIVISIBILITY

8. Factor/Multiple

The **factors** of integer *n* are the positive integers that divide into *n* with no remainder. The **multiples** of *n* are the integers that *n* divides into with no remainder. For example, 6 is a factor of 12, and 24 is a multiple of 12. 12 is both a factor and a multiple of itself, since 12 × 1 = 12 and 12 ÷ 12 = 1.

9. Prime Factorization

To find the prime factorization of an integer, continue factoring until **all the factors are prime**. For example, to factor 36: 36 = 4 × 9 = 2 × 2 × 3 × 3.

10. Relative Primes

Relative primes are integers that have no common factor other than 1. To determine whether two integers are relative primes, break them both down to their prime factorizations. For example, 35 = 5 × 7, and 54 = 2 × 3 × 3 × 3. They have **no prime factors in common**, so 35 and 54 are relative primes.

11. Common Multiple

A common multiple of two or more integers is a number that is a multiple of all of these integers. You can always get a common multiple of two integers by **multiplying** them, but, unless the two numbers are relative primes, the product will not be the *least* common multiple. For example, to find a common multiple for 12 and 15, you could just multiply: 12 × 15 = 180.

To find the **least common multiple** (LCM), test the **multiples of the larger integer** until you find one that's **also a multiple of the smaller**. To find the LCM of 12 and 15, begin by taking the multiples of 15: 15 is not divisible by 12; 30 is not; nor is 45. But the next multiple of 15, 60, *is* divisible by 12, so it's the LCM.

12. Greatest Common Factor (GCF)

To find the greatest common factor of two or more integers, break down the integers into their prime factorizations and multiply **all the prime factors they have in common**. For example, 36 = 2 × 2 × 3 × 3, and 48 = 2 × 2 × 2 × 2 × 3. These integers have a 2 × 2 and a 3 in common, so the GCF is 2 × 2 × 3 = 12.

13. Even/Odd

To predict whether a sum, difference, or product will be even or odd, just **take simple numbers like 1 and 2 and see what happens**. There are rules—"odd times even is even," for example—but there's no need to memorize them. What happens with one set of numbers generally happens with all similar sets.

14. Multiples of 2 and 4

An integer is divisible by 2 (even) if the **last digit** is even. An integer is divisible by 4 if the **last two digits form a multiple of 4**. The last digit of 562 is 2, which is even, so 562 is a multiple of 2. The last two digits form 62, which is *not* divisible by 4, so 562 is not a multiple of 4. The integer 512, however, is divisible by 4 because the last two digits form 12, which is a multiple of 4.

15. Multiples of 3 and 9

An integer is divisible by 3 if the **sum of its digits is divisible by 3**. An integer is divisible by 9 if the **sum of its digits is divisible by 9**. The sum of the digits in 957 is 21, which is divisible by 3 but not by 9, so 957 is divisible by 3 but not by 9.

16. Multiples of 5 and 10

An integer is divisible by 5 if the **last digit is 5 or 0**. An integer is divisible by 10 if the **last digit is 0**. The last digit of 665 is 5, so 665 is a multiple of 5 but *not* a multiple of 10.

17. Remainders

The remainder is the **whole number left over after division**. 487 is 2 more than 485, which is a multiple of 5, so when 487 is divided by 5, the remainder is 2.

FRACTIONS AND DECIMALS

18. Reducing Fractions

To reduce a fraction to lowest terms, **factor out and cancel** all factors the numerator and denominator have in common.

$$\frac{28}{36} = \frac{4 \times 7}{4 \times 9} = \frac{7}{9}$$

19. Adding/Subtracting Fractions

To add or subtract fractions, first find a **common denominator**, then add or subtract the numerators.

$$\frac{2}{15} + \frac{3}{10} = \frac{4}{30} + \frac{9}{30} = \frac{4+9}{30} = \frac{13}{30}$$

20. Multiplying Fractions

To multiply fractions, **multiply** the numerators and **multiply** the denominators.

$$\frac{5}{7} \times \frac{3}{4} = \frac{5 \times 3}{7 \times 4} = \frac{15}{28}$$

21. Dividing Fractions

To divide fractions, **invert** the second one and **multiply**.

$$\frac{1}{2} \div \frac{3}{5} = \frac{1}{2} \times \frac{5}{3} = \frac{1 \times 5}{2 \times 3} = \frac{5}{6}$$

22. Mixed Numbers and Improper Fractions

To convert a mixed number to an improper fraction, **multiply** the whole number part by the denominator, then **add** the numerator. The result is the new numerator (over the same denominator). To convert $7\frac{1}{3}$, first multiply 7 by 3, then add 1, to get the new numerator of 22. Put that over the same denominator, 3, to get $\frac{22}{3}$.

To convert an improper fraction to a mixed number, divide the denominator into the numerator to get a **whole number quotient with a remainder**. The quotient becomes the whole number part of the mixed number, and the remainder becomes the new numerator—with the same denominator. For example, to convert $\frac{108}{5}$, first divide 5 into 108, which yields 21 with a remainder of 3. Therefore, $\frac{108}{5} = 21\frac{3}{5}$.

23. Reciprocal

To find the reciprocal of a fraction, **switch the numerator and the denominator**. The reciprocal of $\frac{3}{7}$ is $\frac{7}{3}$. The reciprocal of 5 is $\frac{1}{5}$. The product of reciprocals is 1.

24. Comparing Fractions

One way to compare fractions is to **re-express them with a common denominator**. $\frac{3}{4} = \frac{21}{28}$ and $\frac{5}{7} = \frac{20}{28}$. $\frac{21}{28}$ is greater than $\frac{20}{28}$, so $\frac{3}{4}$ is greater than $\frac{5}{7}$. Another method is to **convert them both to decimals**: $\frac{3}{4}$ converts to 0.75, and $\frac{5}{7}$ converts to approximately 0.714.

25. Converting Fractions and Decimals

To convert a fraction to a decimal, **divide the bottom into the top**. To convert $\frac{5}{8}$, divide 8 into 5, yielding 0.625.

To convert a decimal to a fraction, set the decimal over 1 and **multiply the numerator and denominator by 10** raised to the number of digits which are to the right of the decimal point.

To convert 0.625 to a fraction, you would multiply $\frac{0.625}{1}$ by $\frac{10^3}{10^3}$ or $\frac{1,000}{1,000}$.

Then simplify: $\frac{625}{1,000} = \frac{5 \times 125}{8 \times 125} = \frac{5}{8}$.

26. Repeating Decimal

To find a particular digit in a repeating decimal, note the **number of digits in the cluster that repeats**. If there are 2 digits in that cluster, then every second digit is the same. If there are 3 digits in that cluster, then every third digit is the same. And so on. For example, the decimal equivalent of $\frac{1}{27}$ is 0.037037037..., which is best written $0.\overline{037}$. There are 3 digits in the repeating cluster, so every third digit is the same. To find the 50th digit, look for the multiple of 3 just less than 50—that's 48. The 48th digit is 7, and with the 49th digit, the pattern repeats with 0. The 50th digit is 3.

27. Identifying the Parts and the Whole

The key to solving most fraction and percent word problems is to identify the **part** and the **whole**. Usually you'll find the **part** associated with the verb *is/are* and the **whole** associated with the word *of*. In the sentence "Half of the boys are blonds," the whole is the boys ("*of* the boys"), and the part is the blonds ("*are* blonds").

PERCENTS

28. Percent Formula

Whether you need to find the part, the whole, or the percent, use the same formula:

$$\textbf{Part} = \textbf{Percent} \times \textbf{Whole}$$

Example: What is 12 percent of 25?
Setup: Part = 0.12 × 25.

Example: 15 is 3 percent of what number?
Setup: 15 = 0.03 × Whole.

Example: 45 is what percent of 9?
Setup: 45 = Percent × 9.

29. Percent Increase and Decrease

To increase a number by a percent, **add the percent to 100 percent**, convert to a decimal, and multiply. To increase 40 by 25 percent, add 25 percent to 100 percent, convert 125 percent to 1.25, and multiply by 40. 1.25 × 40 = 50.

30. Finding the Original Whole

To find the **original whole before a percent increase or decrease**, set up an equation. Think of the result of a 15 percent increase over *x* as 1.15*x*.

Example: After a 5 percent increase, the population was 59,346. What was the population before the increase?
Setup: 1.05*x* = 59,346

31. Combined Percent Increase and Decrease

To determine the combined effect of multiple percent increases and/or decreases, **start with 100 and see what happens**.

Example: A price went up 10 percent one year, and the new price went up 20 percent the next year. What was the combined percent increase?
Setup: First year: 100 + (10 percent of 100) = 110. Second year: 110 + (20 percent of 110) = 132. That's a combined 32 percent increase.

RATIOS, PROPORTIONS, AND RATES

32. Setting Up a Ratio

To find a ratio, put the number associated with the word **of on top** and the quantity associated with the word **to on the bottom** and reduce. The ratio of 20 oranges to 12 apples is $\frac{20}{12}$, which reduces to $\frac{5}{3}$.

33. Part-to-Part Ratios and Part-to-Whole Ratios

If the parts add up to the whole, a part-to-part ratio can be turned into two part-to-whole ratios by putting **each number in the original ratio over the sum of the numbers**. If the ratio of males to females is 1 to 2, then the males-to-people ratio is $\frac{1}{1+2} = \frac{1}{3}$, and the females-to-people ratio is $\frac{2}{1+2} = \frac{2}{3}$. In other words, $\frac{2}{3}$ of all the people are female.

34. Solving a Proportion

To solve a proportion, cross multiply:

$$\frac{x}{5} = \frac{3}{4}$$

$$4x = 3 \times 5$$

$$x = \frac{15}{4} = 3.75$$

35. Rate

To solve a rate problem, **use the units** to keep things straight.

Example: If snow is falling at the rate of one foot every four hours, how many inches of snow will fall in seven hours?

Setup:
$$\frac{1 \text{ foot}}{4 \text{ hours}} = \frac{x \text{ inches}}{7 \text{ hours}}$$

$$\frac{12 \text{ inches}}{4 \text{ hours}} = \frac{x \text{ inches}}{7 \text{ hours}}$$

$$4x = 12 \times 7$$

$$x = 21$$

36. Average Rate

Average rate is *not* simply the average of the rates.

$$\text{Average } A \text{ per } B = \frac{\text{Total } A}{\text{Total } B}$$

$$\text{Average Speed} = \frac{\text{Total distance}}{\text{Total time}}$$

To find the average speed for 120 miles at 40 mph and 120 miles at 60 mph, **don't just average the two speeds**. First, figure out the total distance and the total time. The total distance is 120 + 120 = 240 miles. The times are 3 hours for the first leg and 2 hours for the second leg, or 5 hours total. The average speed, then, is $\frac{240}{5}$ = 48 miles per hour.

AVERAGES

37. Average Formula

To find the average of a set of numbers, **add them up and divide by the number of numbers**.

$$\text{Average} = \frac{\text{Sum of the terms}}{\text{Number of terms}}$$

To find the average of the 5 numbers 12, 15, 23, 40, and 40, first add them: 12 + 15 + 23 + 40 + 40 = 130. Then, divide the sum by 5: 130 ÷ 5 = 26.

38. Average of Evenly Spaced Numbers

To find the average of evenly spaced numbers, just **average the smallest and the largest**. The average of all the integers from 13 through 77 is the same as the average of 13 and 77:

$$\frac{13+77}{2} = \frac{90}{2} = 45$$

39. Using the Average to Find the Sum

$$\text{Sum} = (\text{Average}) \times (\text{Number of terms})$$

If the average of 10 numbers is 50, then they add up to 10 × 50, or 500.

40. Finding the Missing Number

To find a missing number when you're given the average, **use the sum**. If the average of 4 numbers is 7, then the sum of those 4 numbers is 4 × 7, or 28. Suppose that 3 of the numbers are 3, 5, and 8. These 3 numbers add up to 16 of that 28, which leaves 12 for the fourth number.

41. Median and Mode

The median of a set of numbers is the **value that falls in the middle of the set**. If you have 5 test scores and they are 88, 86, 57, 94, and 73, you must first list the scores in increasing or decreasing order: 57, 73, 86, 88, 94.

The median is the middle number, or 86. If there is an even number of values in a set (6 test scores, for instance), simply take the average of the two middle numbers.

The mode of a set of numbers is the **value that appears most often**. If your test scores were 88, 57, 68, 85, 99, 93, 93, 84, and 81, the mode of the scores would be 93 because it appears more often than any other score. If there is a tie for the most common value in a set, the set has more than one mode.

POSSIBILITIES AND PROBABILITY

42. Counting the Possibilities

The fundamental counting principle: If there are **m ways** one event can happen and **n ways** a second event can happen, then there are **$m \times n$ ways** for the two events to happen. For example, with 5 shirts and 7 pairs of pants to choose from, you can have $5 \times 7 = 35$ different outfits.

43. Probability

$$\text{Probability} = \frac{\text{Favorable Outcomes}}{\text{Total Possible Outcomes}}$$

For example, if you have 12 shirts in a drawer and 9 of them are white, the probability of picking a white shirt at random is $\frac{9}{12} = \frac{3}{4}$. This probability can also be expressed as 0.75 or 75%.

POWERS AND ROOTS

44. Multiplying and Dividing Powers

To multiply powers with the same base, **add the exponents and keep the same base:**

$$x^3 \times x^4 = x^{3+4} = x^7$$

To divide powers with the same base, **subtract the exponents and keep the same base:**

$$y^{13} \div y^8 = y^{13-8} = y^5$$

45. Raising Powers to Powers

To raise a power to a power, **multiply the exponents:**

$$(x^3)^4 = x^{3\times4} = x^{12}$$

46. Simplifying Square Roots

To simplify a square root, **factor out the perfect squares** under the radical, unsquare them, and put the result in front.

$$\sqrt{12} = \sqrt{4\times3} = \sqrt{4} \times \sqrt{3} = 2\sqrt{3}$$

47. Adding and Subtracting Roots

You can add or subtract radical expressions **when the part under the radicals is the same:**

$$2\sqrt{3} + 3\sqrt{3} = 5\sqrt{3}$$

Don't try to add or subtract when the radical parts are different. There's not much you can do with an expression like:

$$3\sqrt{5} + 3\sqrt{7}$$

48. Multiplying and Dividing Roots

The product of square roots is equal to the **square root of the product:**

$$\sqrt{3} \times \sqrt{5} = \sqrt{3 \times 5} = \sqrt{15}$$

The quotient of square roots is equal to the **square root of the quotient:**

$$\frac{\sqrt{6}}{\sqrt{3}} = \sqrt{\frac{6}{3}} = \sqrt{2}$$

49. Negative Exponent and Rational Exponent

To find the value of a number raised to a negative exponent, simply rewrite the number, without the negative sign, as the bottom of a fraction with 1 as the numerator of the fraction: $3^{-2} = \frac{1}{3^2} = \frac{1}{9}$. If x is a positive number and a is a nonzero number, then $x^{\frac{1}{a}} = \sqrt[a]{x}$. So $4^{\frac{1}{2}} = \sqrt[2]{4} = 2$. If p and q are integers, then $x^{\frac{p}{q}} = \sqrt[q]{x^p}$. So $4^{\frac{3}{2}} = \sqrt[2]{4^3} = \sqrt{64} = 8$.

ABSOLUTE VALUE

50. Determining Absolute Value

The absolute value of a number is the distance of the number from zero on the number line. Because absolute value is a distance, it is always positive. The absolute value of 7 is 7; this is expressed $|7| = 7$. Similarly, the absolute value of –7 is 7: $|-7| = 7$. Every positive number is the absolute value of two numbers: itself and its negative.

ALGEBRAIC EXPRESSIONS

51. Evaluating an Expression

To evaluate an algebraic expression, **plug in** the given values for the unknowns and calculate according to **PEMDAS**. To find the value of $x^2 + 5x - 6$ when $x = -2$, plug in -2 for x: $(-2)^2 + 5(-2) - 6 = -12$.

52. Adding and Subtracting Monomials

To combine like terms, **keep the variable part unchanged while adding or subtracting the coefficients:**

$$2a + 3a = (2 + 3)a = 5a$$

53. Adding and Subtracting Polynomials

To add or subtract polynomials, **combine like terms**.

$$(3x^2 + 5x - 7) - (x^2 + 12) =$$
$$(3x^2 - x^2) + 5x + (-7 - 12) =$$
$$2x^2 + 5x - 19$$

54. Multiplying Monomials

To multiply monomials, **multiply the coefficients and the variables separately:**

$$2a \times 3a = (2 \times 3)(a \times a) = 6a^2$$

55. Multiplying Binomials—FOIL

To multiply binomials, use **FOIL**. To multiply $(x + 3)$ by $(x + 4)$, first multiply the **F**irst terms: $x \times x = x^2$. Next the **O**uter terms: $x \times 4 = 4x$. Then the **I**nner terms: $3 \times x = 3x$. And finally the **L**ast terms: $3 \times 4 = 12$. Then add and combine like terms:

$$x^2 + 4x + 3x + 12 = x^2 + 7x + 12$$

56. Multiplying Other Polynomials

FOIL works only when you want to multiply two binomials. If you want to multiply polynomials with more than two terms, make sure you **multiply each term in the first polynomial by each term in the second**.

$$(x^2 + 3x + 4)(x + 5) =$$
$$x^2(x + 5) + 3x(x + 5) + 4(x + 5) =$$
$$x^3 + 5x^2 + 3x^2 + 15x + 4x + 20 =$$
$$x^3 + 8x^2 + 19x + 20$$

After multiplying two polynomials together, the number of terms in your expression before simplifying should equal the number of terms in one polynomial multiplied by the number of terms in the second. In the example, you should have $3 \times 2 = 6$ terms in the product before you simplify like terms.

FACTORING ALGEBRAIC EXPRESSIONS

57. Factoring out a Common Divisor

A factor common to all terms of a polynomial can be **factored out**. All three terms in the polynomial $3x^3 + 12x^2 - 6x$ contain a factor of $3x$. Pulling out the common factor yields $3x(x^2 + 4x - 2)$.

58. Factoring the Difference of Squares

One of the test maker's favorite factorables is the **difference of squares**.

$$a^2 - b^2 = (a - b)(a + b)$$

$x^2 - 9$, for example, factors to $(x - 3)(x + 3)$.

59. Factoring the Square of a Binomial

Recognize polynomials that are squares of binomials:

$$a^2 + 2ab + b^2 = (a + b)^2$$
$$a^2 - 2ab + b^2 = (a - b)^2$$

For example, $4x^2 + 12x + 9$ factors to $(2x + 3)^2$, and $n^2 - 10n + 25$ factors to $(n - 5)^2$.

60. Factoring Other Polynomials—FOIL in Reverse

To factor a quadratic expression, **think about what binomials you could use FOIL on to get that quadratic expression**. To factor $x^2 - 5x + 6$, think about what **F**irst terms will produce x^2, what **L**ast terms will produce $+6$, and what **O**uter and **I**nner terms will produce $-5x$. Some common sense—and a little trial and error—lead you to $(x - 2)(x - 3)$.

61. Simplifying an Algebraic Fraction

Simplifying an algebraic fraction is a lot like simplifying a numerical fraction. The general idea is to **find factors common to the numerator and denominator and cancel them**. Thus, simplifying an algebraic fraction begins with factoring.

For example, to simplify $\dfrac{x^2 - x - 12}{x^2 - 9}$, first factor the numerator and denominator:

$$\frac{x^2 - x - 12}{x^2 - 9} = \frac{(x-4)(x+3)}{(x-3)(x+3)}$$

Canceling $x + 3$ from the numerator and denominator leaves you with $\dfrac{x-4}{x-3}$.

SOLVING EQUATIONS

62. Solving a Linear Equation

To solve an equation, do whatever is necessary to both sides to **isolate the variable**. To solve the equation $5x - 12 = -2x + 9$, first get all the x's on one side by adding $2x$ to both sides: $7x - 12 = 9$. Then add 12 to both sides: $7x = 21$. Then divide both sides by 7: $x = 3$.

63. Solving "In Terms Of"

To solve an equation for one variable **in terms of** another means to **isolate the one variable on one side of the equation**, leaving an expression containing the other variable on the other side of the equation. To solve the equation $3x - 10y = -5x + 6y$ for x in terms of y, isolate x:

$$3x - 10y = -5x + 6y$$
$$3x + 5x = 6y + 10y$$
$$8x = 16y$$
$$x = 2y$$

64. Translating from English into Algebra

To translate from English into algebra, look for the key words and systematically turn phrases into algebraic expressions and sentences into equations. Be careful about order, especially when subtraction is called for.

Example: Celine and Remi play tennis. Last year, Celine won 3 more than twice the number of matches that Remi won. If Celine won 11 more matches than Remi, how many matches did Celine win?

Setup: You are given two sets of information. One way to solve this is to write a system of equations–one equation for each set of information. Use variables that relate well with what they represent. For example, use r to represent Remi's winning matches. Use c to represent Celine's winning matches. The phrase "Celine won 3 more than twice Remi" can be written as $c = 2r + 3$. The phrase "Celine won 11 more matches than Remi" can be written as $c = r + 11$.

65. Solving a Quadratic Equation

To solve a quadratic equation, put it in the "$ax^2 + bx + c = 0$" form, **factor** the left side (if you can), and set each factor equal to 0 separately to get the two solutions. To solve $x^2 + 12 = 7x$, first rewrite it as $x^2 - 7x + 12 = 0$. Then factor the left side:

$$(x - 3)(x - 4) = 0$$
$$x - 3 = 0 \text{ or } x - 4 = 0$$
$$x = 3 \text{ or } 4$$

66. Solving a System of Equations

You can solve for two variables only if you have two distinct equations. Two forms of the same equation will not be adequate. **Combine the equations** in such a way that **one of the variables cancels out**. To solve the two equations $4x + 3y = 8$ and $x + y = 3$, multiply both sides of the second equation by -3 to get: $-3x - 3y = -9$. Now add the two equations; the $3y$ and the $-3y$ cancel out, leaving $x = -1$. Plug that back into either one of the original equations, and you'll find that $y = 4$.

67. Solving an Inequality

To solve an inequality, do whatever is necessary to both sides to **isolate the variable**. Just remember that when you **multiply or divide both sides by a negative number**, you must **reverse the sign**. To solve −5x + 7 < −3, subtract 7 from both sides to get −5x < −10. Now divide both sides by −5, remembering to reverse the sign: x > 2.

68. Radical Equations

A radical equation is one that contains at least one radical expression. Solve radical equations by using standard rules of algebra. If $5\sqrt{x} - 2 = 13$, then $5\sqrt{x} = 15$ and $\sqrt{x} = 3$, so x = 9.

FUNCTIONS

69. Function Notation and Evaluation

Standard function notation is written $f(x)$ and read "f of x." To evaluate the function $f(x) = 2x + 3$ for $f(4)$, replace x with 4 and simplify: $f(4) = 2(4) + 3 = 11$.

70. Direct and Inverse Variation

In direct variation, $y = kx$, where k is a nonzero constant. In direct variation, the variable y changes directly as x does. If a unit of Currency A is worth 2 units of Currency B, then A = 2B. If the number of units of B were to double, the number of units of A would double, and so on for halving, tripling, etc. In inverse variation, $xy = k$, where x and y are variables and k is a constant. A famous inverse relationship is rate × time = distance, where distance is constant. Imagine having to cover a distance of 24 miles. If you were to travel at 12 miles per hour, you'd need 2 hours. But if you were to halve your rate, you would have to double your time. This is just another way of saying that rate and time vary inversely.

71. Domain and Range of a Function

The domain of a function is the set of values for which the function is defined. For example, the domain of $f(x) = \dfrac{1}{1 - x^2}$ is all values of x except 1 and −1, because for those values the denominator has a value of 0 and the fraction is therefore undefined. The range of a function is the set of outputs or results of the function. For example, the range of $f(x) = x^2$ is all numbers greater than or equal to zero, because x^2 cannot be negative.

COORDINATE GEOMETRY

72. Finding the Distance Between Two Points

To find the distance between points, **use the Pythagorean theorem** or **special right triangles**. The difference between the x's is one leg and the difference between the y's is the other.

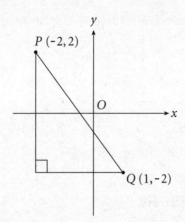

In the figure above, *PQ* is the hypotenuse of a 3-4-5 triangle, so *PQ* = 5.
You can also use the **distance formula**:

$$d = \sqrt{(x_1 - x_2)^2 + (y_1 - y_2)^2}$$

To find the distance between *R*(3,6) and *S*(5,–2):

$$d = \sqrt{(3-5)^2 + [6-(-2)]^2}$$
$$= \sqrt{(-2)^2 + (8)^2}$$
$$= \sqrt{68} = 2\sqrt{17}$$

73. Using Two Points to Find the Slope

$$\text{Slope} = \frac{\text{Change in } y}{\text{Change in } x} = \frac{\text{Rise}}{\text{Run}}$$

The slope of the line that contains the points *A*(2,3) and *B*(0,–1) is

$$\frac{y_1 - y_2}{x_1 - x_2} = \frac{3-(-1)}{2-0} = \frac{4}{2} = 2$$

74. Using an Equation to Find the Slope

To find the slope of a line from an equation, put the equation into the **slope-intercept** form:

$$y = mx + b$$

The **slope is *m***. To find the slope of the equation $3x + 2y = 4$, rearrange it:

$$3x + 2y = 4$$
$$2y = -3x + 4$$
$$y = -\frac{3}{2}x + 2$$

The slope is $-\frac{3}{2}$.

75. Using an Equation to Find an Intercept

To find the *y*-intercept, you can either put the equation into ***y* = *mx* + *b* (slope-intercept)** form—in which case ***b* is the *y*-intercept**—or you can just **plug *x* = 0** into the equation and **solve for *y***. To find the *x*-intercept, plug ***y* = 0** into the equation and **solve for *x***.

76. Finding the Midpoint

The midpoint of two points on a line segment is the average of the *x*-coordinates of the endpoints and the average of the *y*-coordinates of the endpoints. If the endpoints are (x_1, y_1) and (x_2, y_2), the midpoint is $\left(\dfrac{x_1 + x_2}{2}, \dfrac{y_1 + y_2}{2}\right)$. The midpoint of (3,5) and (9,1) is $\left(\dfrac{3+9}{2}, \dfrac{5+1}{2}\right)$ or (6, 3).

LINES AND ANGLES

77. Intersecting Lines

When two lines intersect, **adjacent angles are supplementary, and vertical angles are equal**.

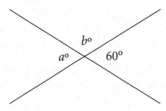

In the figure above, the angles marked *a*° and *b*° are adjacent and supplementary, so $a + b = 180$. Furthermore, the angles marked *a*° and 60° are vertical and equal, so $a = 60$.

78. Parallel Lines and Transversals

A transversal across parallel lines forms **four equal acute angles and four equal obtuse angles**, unless the transversal meets the lines at a right angle; then all eight angles are right angles.

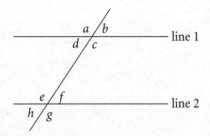

In the figure above, line 1 is parallel to line 2. Angles *a*, *c*, *e*, and *g* are obtuse, so they are all equal. Angles *b*, *d*, *f*, and *h* are acute, so they are all equal.

Furthermore, **any of the acute angles is supplementary to any of the obtuse angles**. Angles *a* and *h* are supplementary, as are *b* and *e*, *c* and *f*, and so on.

TRIANGLES—GENERAL

79. Interior and Exterior Angles of a Triangle

The three angles of any triangle **add up to 180 degrees**.

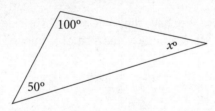

In the figure above, $x + 50 + 100 = 180$, so $x = 30$.

An exterior angle of a triangle is equal to the **sum of the remote interior angles**.

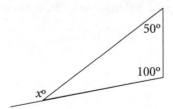

In the figure above, the exterior angle labeled $x°$ is equal to the sum of the remote angles: $x = 50 + 100 = 150$.

The three exterior angles of a triangle **add up to 360 degrees**.

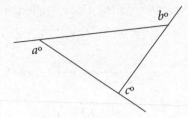

In the figure above, $a + b + c = 360$.

80. Similar Triangles

Similar triangles have the same shape: **corresponding angles are equal and corresponding sides are proportional**.

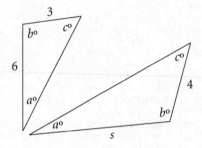

The triangles above are similar because they have the same angles. The side of length 3 corresponds to the side of length 4, and the side of length 6 corresponds to the side of length s.

$$\frac{3}{4} = \frac{6}{s}$$
$$3s = 24$$
$$s = 8$$

81. Area of a Triangle

$$\text{Area of Triangle} = \frac{1}{2}\,(\text{base})(\text{height})$$

The height is the perpendicular distance between the side that's chosen as the base and the opposite vertex.

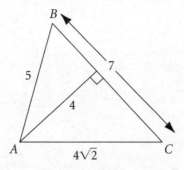

In the triangle above, 4 is the height when 7 is chosen as the base.

$$\text{Area} = \frac{1}{2}bh = \frac{1}{2}(7)(4) = 14$$

82. Triangle Inequality Theorem

The length of one side of a triangle must be **greater than the difference between and less than the sum** of the lengths of the other two sides. For example, if it is given that the length of one side is 3 and the length of another side is 7, then you know that the length of the third side must be greater than 7 − 3 = 4 and less than 7 + 3 = 10.

83. Isosceles and Equilateral Triangles

An isosceles triangle is a triangle that has **two equal sides**. Not only are two sides equal, but the angles opposite the equal sides, called **base angles**, are also equal.

Equilateral triangles are triangles in which **all three sides are equal**. Since all the sides are equal, all the angles are also equal. All three angles in an equilateral triangle measure 60 degrees, regardless of the lengths of the sides.

RIGHT TRIANGLES

84. Pythagorean Theorem

For all right triangles:

$$(\text{leg}_1)^2 + (\text{leg}_2)^2 = (\text{hypotenuse})^2$$

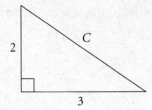

If one leg is 2 and the other leg is 3, then:

$$2^2 + 3^2 = c^2$$
$$c^2 = 4 + 9$$
$$c = \sqrt{13}$$

85. The 3-4-5 Triangle

If a right triangle's leg-to-leg ratio is 3:4, or if the leg-to-hypotenuse ratio is 3:5 or 4:5, it's a 3-4-5 triangle, and you don't need to use the Pythagorean theorem to find the third side. Just figure out what multiple of 3-4-5 it is.

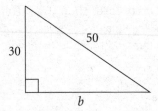

In the right triangle shown, one leg is 30, and the hypotenuse is 50. This is 10 times 3-4-5. The other leg is 40.

86. The 5-12-13 Triangle

If a right triangle's leg-to-leg ratio is 5:12, or if the leg-to-hypotenuse ratio is 5:13 or 12:13, then it's a 5-12-13 triangle, and you don't need to use the Pythagorean theorem to find the third side. Just figure out what multiple of 5-12-13 it is.

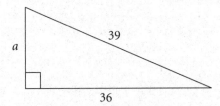

Here one leg is 36, and the hypotenuse is 39. This is 3 times 5-12-13. The other leg is 15.

87. The 30-60-90 Triangle

The sides of a 30-60-90 triangle are in a ratio of $x : x\sqrt{3} : 2x$. You don't need the Pythagorean theorem.

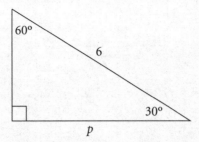

If the hypotenuse is 6, then the shorter leg is half that, or 3, and then the longer leg is equal to the short leg times $\sqrt{3}$, or $p = 3\sqrt{3}$.

88. The 45-45-90 Triangle

The sides of a 45-45-90 triangle are in a ratio of $x : x : x\sqrt{2}$.

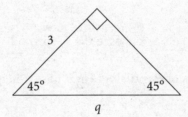

If one leg has a length of 3, then the other leg also has a length of 3, and the hypotenuse is equal to a leg times $\sqrt{2}$, or $q = 3\sqrt{2}$.

OTHER POLYGONS

89. Characteristics of a Rectangle

A rectangle is a **four-sided figure with four right angles**. Opposite sides are equal. Diagonals are equal.

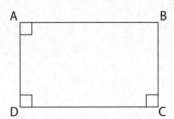

Quadrilateral *ABCD* above is shown to have three right angles. The fourth angle therefore also measures 90 degrees, and *ABCD* is a rectangle. The **perimeter** of a rectangle is equal to the sum of the lengths of the four sides, which is equivalent to 2(length + width).

Area of Rectangle = length × width

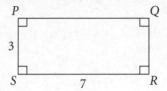

The area of a 7-by-3 rectangle is $7 \times 3 = 21$.

90. Characteristics of a Parallelogram

A parallelogram has **two pairs of parallel sides**. Opposite sides are equal. Opposite angles are equal. Consecutive angles add up to 180 degrees.

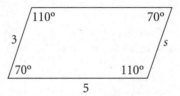

In the figure above, s is the length of the side opposite the 3, so $s = 3$.

Area of Parallelogram = base × height

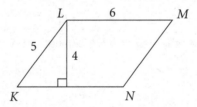

In parallelogram *KLMN* above, 4 is the height when *LM* or *KN* is used as the base.

Base × height = $6 \times 4 = 24$.

91. Characteristics of a Square

A square is a **rectangle with four equal sides**.

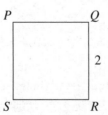

If *PQRS* is a square, all sides are the same length as *QR*. The **perimeter** of a square is equal to four times the length of one side.

$$\text{Area of Square} = (\text{side})^2$$

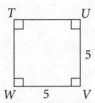

The square above, with sides of length 5, has an area of $5^2 = 25$.

92. Interior Angles of a Polygon

The **sum of the measures of the interior angles of a polygon = $(n - 2) \times 180$**, where *n* is the number of sides.

$$\text{Sum of the Angles} = (n - 2) \times 180$$

The eight angles of an octagon, for example, add up to $(8 - 2) \times 180 = 1{,}080$.

CIRCLES

93. Circumference of a Circle

$$\text{Circumference} = 2\pi r$$

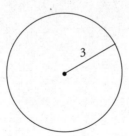

In the circle above, the radius has a length of 3, so the circumference is $2\pi(3) = 6\pi$.

94. Length of an Arc

An arc is a piece of the circumference. If *n* is the degree measure of the arc's central angle, then the formula is

$$\text{Length of an Arc} = \left(\frac{n}{360}\right)(2\pi r)$$

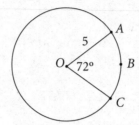

In the figure above, the radius has a length of 5, and the measure of the central angle is 72 degrees. The arc length is $\frac{72}{360}$, or $\frac{1}{5}$, of the circumference:

$$\left(\frac{72}{360}\right)(2\pi)(5) = \left(\frac{1}{5}\right)(10\pi) = 2\pi$$

95. Area of a Circle

$$\text{Area of a Circle} = \pi r^2$$

The area of the circle is $\pi(4)^2 = 16\pi$.

96. Area of a Sector

A sector is a piece of the area of a circle. If *n* is the degree measure of the sector's central angle, then the formula is

$$\text{Area of a Sector} = \left(\frac{n}{360}\right)(\pi r^2)$$

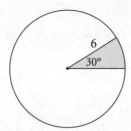

In the figure above, the radius has a length of 6, and the measure of the sector's central angle is 30 degrees. The sector has $\frac{30}{360}$, or $\frac{1}{12}$, of the area of the circle:

$$\left(\frac{30}{360}\right)(\pi)(6^2)=\left(\frac{1}{12}\right)(36\pi)=3\pi$$

97. Tangency

When a line is tangent to a circle, the radius of the circle is perpendicular to the line at the point of contact.

SOLIDS

98. Surface Area of a Rectangular Solid

The surface of a rectangular solid consists of three pairs of identical faces. To find the surface area, find the area of each face and add them up. If the length is *l*, the width is *w*, and the height is *h*, the formula is

Surface Area = 2*lw* + 2*wh* + 2*lh*

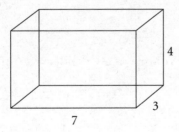

The surface area of the box above is: $(2 \times 7 \times 3) + (2 \times 3 \times 4) + (2 \times 7 \times 4) = 42 + 24 + 56 = 122$

99. Volume of a Rectangular Solid

Volume of a Rectangular Solid = *lwh*

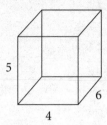

The volume of a 4-by-5-by-6 box is

$$4 \times 5 \times 6 = 120$$

A cube is a rectangular solid with length, width, and height all equal. If *s* is the length of an edge of a cube, the volume formula is

Volume of a Cube = *s*³

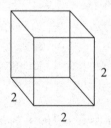

The volume of this cube is $2^3 = 8$.

100. Volume of a Cylinder

Volume of a Cylinder = *πr²h*

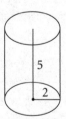

In the cylinder above, *r* = 2, *h* = 5, so

$$\text{Volume} = \pi(2^2)(5) = 20\pi$$

NOTES

NOTES

NOTES

NOTES